The Forest Guide
WALES

BLOOMSBURY WILDLIFE
Bloomsbury Publishing Plc
50 Bedford Square, London, WC1B 3DP, UK
29 Earlsfort Terrace, Dublin 2, Ireland

BLOOMSBURY, BLOOMSBURY WILDLIFE
and the Diana logo are trademarks of
Bloomsbury Publishing Plc

First published in the United Kingdom 2025

Copyright © Gabriel Hemery, 2025
Photographs © Gabriel Hemery, 2025
Maps © John Plumer, 2025

Gabriel Hemery has asserted his right under the Copyright, Designs and Patents Act, 1988, to be identified as Author of this work.

For legal purposes the Acknowledgements on p. 288 constitute an extension of this copyright page.

All rights reserved. No part of this publication may be reproduced or transmitted in any form or by any means, electronic or mechanical, including photocopying, recording, or any information storage or retrieval system, without prior permission in writing from the publishers.

Bloomsbury Publishing Plc does not have any control over, or responsibility for, any third-party websites referred to or in this book.

All internet addresses given in this book were correct at the time of going to press. The author and publisher regret any inconvenience caused if addresses have changed or sites have ceased to exist, but can accept no responsibility for any such changes.

A catalogue record for this book is available from the British Library

Library of Congress Cataloguing-in-Publication data has been applied for

ISBN: PB: 978-1-3994-0912-4
ePub: 978-1-3994-0913-1
ePDF: 978-1-3994-0910-0

10 9 8 7 6 5 4 3 2 1

Design and typeset by Austin Taylor

Printed and bound in Turkey by Elma Basım

To find out more about our authors and books visit www.bloomsbury.com and sign up for our newsletters.

BOOK PATRONS

Alison Edwards
Ayse Sabuncu
Colin Johnstone
Esmond Harris MBE
Helen Jane Fangman
Ian & Lit Smith
John Philip Leefe
Linda A. Dolata
Peter Barker
Simon Fineman
Westhope

The Forest Guide
WALES

Copses, Woods and Forests

GABRIEL HEMERY

BLOOMSBURY WILDLIFE
LONDON · OXFORD · NEW YORK · NEW DELHI · SYDNEY

Page 1: An extraordinary oak tree at Coedydd Aber [15].
Previous pages: The trees of Ysgyryd Fawr/The Skirrid [293] with Y Fâl/Sugar Loaf in the distance.
This page: Coed y Brenin Forest Park [59].

CONTENTS

Wales' Forests 7 | How to Use This Guide 20 | Access to Forests 25 | Keeping Safe 27

NORTH-WEST – YNYS MÔN/ISLE OF ANGLESEY, CONWY, GWYNEDD (SITES 1–72) 28

NORTH-EAST – DENBIGHSHIRE, FLINTSHIRE, WREXHAM (SITES 73–101) 82

CENTRAL-WEST – CEREDIGION (SITES 102–135) 106

CENTRAL-EAST – POWYS (SITES 136–200) 130

SOUTH-WEST – PEMBROKESHIRE, CARMARTHENSHIRE, SWANSEA, NEATH PORT TALBOT (SITES 201–263) 168

SOUTH-CENTRAL – MERTHYR TYDFIL, CARDIFF, VALE OF GLAMORGAN, BRIDGEND, RHONDDA CYNON TAF (SITES 264–286) 216

SOUTH-EAST – MONMOUTHSHIRE, NEWPORT, CAERPHILLY, BLAENAU GWENT, TORFAEN (SITES 287–325) 234

Site Designations 268 | Cymraeg Terms 269 | Glossary 272 | Useful Information 274
Further Reading 275 | Regional Maps 276 | Index 283 | Acknowledgements 288

WALES' FORESTS

The copses, woods and forests of Wales are gloriously diverse, in nature, landscape, history and utility. The forests of the country have nurtured us, from early civilisation to the Industrial Revolution and beyond, and this history is evident everywhere. Blessed by moist atmospheric conditions and offered protection by dramatic terrain, Wales harbours some of our most precious habitats, including temperate rainforest. Elsewhere, expansive conifer plantations provide spaces for us to play and explore, while contributing to our timber needs. Add in hidden dingles, tiny copses and clifftop woods, and it is easy to understand why the forestscapes of Wales offer so much to those who love nature and exploring the outdoors.

With so many potential forests to choose from, and given their incredible diversity, selecting the sites to include in this guide was never going to be an easy task.

Dog violet.

Opposite: A secret waterfall tumbles through Coed Maenarthur [124].

Ultimately, 325 sites were chosen, spread right across Wales, offering something of interest to everyone, whether you are touring by car or using public transport, hiking or cycling. Some are urban forests, or located next to car parks offering full facilities; others will require careful planning and experience in exploring demanding terrain. With this unique selection, you will never find yourself far away from a forest site, every one of them offering an unforgettable experience. This guide will help you find Wales' most beautiful, intriguing and wildlife-rich treescapes.

Forest facts and figures

Britain is one of the least-forested places in Europe, with just 13 per cent of its land area covered by forests (only Ireland and the Netherlands have less). Forest cover in Wales is 15 per cent, and so above average for Britain, with Scotland having 19 per cent and England 10 per cent.

What is a hectare (ha)?

A hectare is 10,000m² or 100×100m, and equivalent to 2.47 acres.

Britain has an unusually high degree of private forest ownership compared to European countries, with almost three-quarters of its forest area owned by private individuals, companies and non-governmental organisations. In Wales, 63 per cent of forest area is privately owned, the remaining proportion in public hands, managed on behalf of British citizens by a range of public bodies, notably Natural Resources Wales, the Crown Estate and others, including local authorities. See also Forest ownership (p.17).

Proportions of coniferous and broadleaved forests are closely balanced, with 45 per cent of forest area coniferous within Wales, unlike in Scotland, where almost three-quarters of its forest area is under conifers. Within publicly owned land in Wales, 64 per cent of the forest area is coniferous, while in private forests it is broadleaves that cover the majority of the area (85 per cent).

Central-West Wales is the most forested region, with 18 per cent cover, followed by South-Central (16 per cent). Both two northern regions have 11 per cent forest cover, meaning they are below average for Wales.

Notable sites	
Largest	Wales has many very large conifer plantations. Tywi Forest [127] covers 6,112ha in the Central-West region, while Dyfi Forest [70] extends across 6,000ha in the North-West region.
Smallest	Within this guide, 11 sites are 1ha or less in area, many of them precious habitats managed by local Wildlife Trusts, such as Cors Bodgynydd [29] and Rogiet Poorland [325].
Curious names	Forest Wood [205] might seem to have one name too many. Rogiet Poorland [325] was aptly named. Coedlan Llanfairpwll [18] grows near a village with Britain's longest placename Llanfairpwllgwyngyllgogerychwyrndrobwllllantysiliogogogoch, which literally means: 'the church of St Mary of the pool of the white hazels (*gwyn gyll*) near to the fierce whirlpool and the church of St Tysilio of the red cave'.
Most remote	Areas of Tywi Forest [127] are very remote. The Nant Syddion bothy (one of only nine in Wales) hidden within Myherin Forest [110] provides the opportunity for an unforgettable (perhaps spooky) night in the forest.
Region with most temperate rainforest sites	With 21 sites, the North-West has almost two-thirds of the 36 temperate rainforest sites featured in this guide.
Tallest trees	Several sites have very tall Douglas fir (60m plus), including Gwydir Forest [34] and Lake Vyrnwy [138]. Also a giant redwood at Bodnant Garden [12], which exceeds 50m.
Rarest trees	Several very rare 'micro-species' of *Sorbus* are found in Wales, including the Menai whitebeam at Nantporth [14], English whitebeam at Piercefield Woods [318] and Ley's whitebeam at Darren Fach [265].
Compass extremities	North: Lawr yr Afon [1] (North-West) on the Ynys Môn/Isle of Anglesey. East: Highmeadow Woods [298] (South-East), which is accessed by walking from England. South: Cwm Colhuw [286] (South-Central). West: Porth Clais [228] (South-West) on St Davids Peninsula.
Greatest density of sites by OS 1:25,000 map	OS Map 213 Aberystwyth & Cwm Rheidol has 19 sites, followed by OL14 Wye Valley & Forest of Dean, with 16 sites.
Most forested region	Central-West Wales is proportionally the most forested region, with 18 per cent forest cover (approaching double the average for England). Central-East has the greatest forested area, with 72,211ha.

Geography

Wales is commonly divided into six regions by the Welsh Government, but in this guide the extensive Mid Wales, or Central Wales, region is divided into two, with the counties of Ceredigion forming Central-West and Powys forming Central-East, creating seven regions in total.

Wales is 20,779km² (8,023 square miles), an area often used as a metric in its own right, particularly in the context of defining areas of tropical rainforest destroyed annually elsewhere in the world. Wales has over 2,700km (1,680 miles) of coastline, comprising its borders to the north, west and south. Central-East is the only region that is landlocked.

The country is dominated by a maritime climate, with moist westerly temperate winds blowing in from the Atlantic Ocean. Most of Wales is mountainous, especially the Eryri (Snowdon) range in the North-West, which has 15 peaks above 914m (3,000 feet), while the Mynyddoedd Cambrian (Cambrian Mountains) run north-east to south-west through the centre of the country. Bannau Brycheiniog (Brecon Beacons), Y Mynyddoedd Duon (Black Mountains) and Y Mynydd Du (Black Mountain) are in the south of the country. The rivers Dyfrdwy (Dee) and Hafren (Severn) partly form the boundary between Wales and England, while the Mawddach, Dyfi (Dovey), Rheidol, Ystwyth and Teifi flow westwards into Bae Ceredigion (Cardigan Bay), and the rivers Tywi (Towy), Taf (Taff), Wysg (Usk) and Gwy (Wye) flow south into the Bristol Channel.

The geology of Wales is highly diverse and celebrated, with the work of early geologists defining many terms now commonly adopted in geological studies. Underlying bedrock and sedimentary deposits affect soil pH and moisture regimes, and therefore the plants that grow on them. They can also provide valuable commodities, and indeed, their exploitation in Wales has shaped the landscape. Examples include valuable minerals, such as iron, copper and lead, deposits of coal, and building products such as lime and slate.

There are three national parks in Wales, covering an area of 4,122km². Snowdonia, now known as Eryri, was the first to be designated in 1951, and is a protected landscape covering 2,131km² in the North-West region (representing one-tenth of the area of Wales, it is sometimes referred to as a 'deciwales'). Brecon Beacons, now known as Bannau Brycheiniog, includes 1,340km² of land, mostly in the south of Central-East Wales, although also covering small areas in all three southern regions. The Pembrokeshire National Park (Parc Cenedlaethol Arfordir Penfro in Welsh) protects a coastal strip covering 622km² around the South-West region, with no part further than 16km from the sea.

Formerly known as Areas of Outstanding Natural Beauty (AONBs), there are five

The crags of Near Hearkening Rock in High Meadow Woods [298].

National Landscapes (NL) that protect other landscapes across Wales, covering 4 per cent of the land area. These are Ynys Môn (Isle of Anglesey) and Llŷn Peninsula in the North-West, Bryniau Clwyd a Dyffryn Dyfrdwy (Clwydian Range and Dee Valley) in the North-East, Gower in the South-West, and Wye Valley in the South-East, which also extends into England.

The upper limit of natural tree growth, or the 'treeline', is about 600m above sea level in Wales, although native woods are rare at this altitude due to the pressures of grazing in the uplands. Some conifer plantations reach this altitude in the North-West, but this is near the limit, even for Sitka spruce. The boundary between managed lower altitude and more intensively managed land, and unenclosed hill and moorland is known as 'ffridd' in Wales (as in Ffridd Wood [149]).

Trees and natural history

The nature of Wales is wonderfully diverse, not only in its forests, but also in its rivers and lakes, marshes and along its coasts. Its uplands, however, are a different matter (more on this below). The temperate rainforests (see p.54) and their rich assemblages of bryophytes (see pp.116–17), flowering plants and iconic woodland birds are the crown jewels.

Britain has a relatively small diversity of about 60 native tree species, meaning those species, subspecies or hybrids that have established themselves without the hand of humans. Of these, only 35 are widespread, and only three are conifers: juniper, Scots pine and yew. The remainder are present in tiny populations, and botanists are frequently refining the list, which includes 17 'micro-species' of *Sorbus*. Introduced tree species are sometimes referred to by conservationists, rather derogatively, as 'aliens' or 'exotics', but they can play an important role in helping forests establish in difficult places and support biodiversity that otherwise may be threatened by environmental change. We have even discovered that some non-native exotics can be beneficial for wildlife, for example lodgepole pine, which is a food source for red squirrels (see Cwm Berwyn Forest [130]).

A recent State of Nature Report (2023) for Wales presented the stark news that

Bluebells emerge in springtime at Craig Cerrig Gleisiad [191] in the Bannau Brycheiniog/Brecon Beacons National Park, indicating that ancient woodland once grew here. Grazing by sheep makes it a barren site, even though this site has multiple conservation designations, including IPA, NNR, NP, SAC and SSSI.

Grazing sheep and fast-spreading rhododendron are both threats to native woodland habitat.

wildlife species continue to decline across the country, and following the loss of 73 species in recent years, another 17 per cent of animal species are now at risk. The reasons are manifold, from urban development and changes in land practice, to collectors seeking the eggs of peregrine falcons and other raptors, and plants like the Snowdon lily. Yet there are glimmers of hope, especially in examples such as the red kite, whose population in the early twentieth century fell to about a dozen individuals surviving in west Wales, but which is now thriving across much of Wales and England thanks to a reintroduction programme. More recently, various innovative conservation approaches are seeking to repeat this success by working with nature to restore ecosystems. The Vincent Wildlife Trust has successfully translocated around 50 pine martens from Scotland to Mid Wales, where it is hoped they will help control the invasive grey squirrel population in favour of our native red squirrel.

One of the biggest debates facing Wales is the future of its uplands. The issues discussed are often contentious, with the potential outcomes highly impactful (in ways both good and bad) to nature and society. Upland hill farming is a heritage industry for Wales, but one that is thought to be impoverishing the lands on which it is based. We face both a climate and a biodiversity crisis, yet upland hill farming is supported by government subsidies, and the withdrawal of these subsidies could result in the failure of many farming businesses, which often rely on subsidies for more than half their income. While the loss of upland hill farming would result in the loss of centuries-old agricultural systems, put an end to family dynasties and lead to a change in the open vistas of our uplands, there are those who believe these are consequences that are justifiable. It is clear that Wales has a huge area of land, often known as 'green desert', in its uplands, which remain treeless due to ceaseless grazing by livestock especially with sheep.

One of the most perverse facts about the state of nature in Wales is that there is greater tree cover in the city of London than in any one of its national parks. In total there are 55,624ha of forest cover across the three national parks, equivalent to 13 per cent tree cover, while in the city of London there is 21 per cent tree cover, which translates as a total of 33,159ha (more than half of the total forest cover of all three national parks in Wales).

It is perhaps most difficult to understand why many sites recognised under our highest conservation designations in Wales, including National Nature Reserves and Sites of Special Scientific Interest, remain in terrible condition due to overgrazing. Visit any of these sites and the lack of naturally regenerating trees is evident, with scattered remnants of woodland and isolated trees clinging to the few cliffs and rocky slopes that are out of reach of grazing livestock.

It seems clear that the situation will not be resolved until either the climate or nature emergencies come closer still to home. While these may provide the impetus to motivate change, there is some hope that change will happen sooner, in the form of some attractive incentives developed by innovative organisations. These organisations are working with farmers to help them recognise that there can be real benefit from the economic value in nature itself, while protecting and enhancing the environment is good business. Stump Up For Trees (see pp.240–41) is one such organisation, working with hill farmers from a position of authority rather than preaching to them. Let's hope that government agencies in Wales act for the good of nature, hand in hand with society.

History of Wales' forests

The history of the land of Wales and its people is deeply entwined with trees and forests. Nine thousand years ago, Wales finally emerged from the last Ice Age, after cyclical periods of cold and warm which saw the ice sheet grow and shrink repeatedly since its maximum about 18,000 years ago. As the new land was exposed and the climate warmed, both trees and people moved in, almost simultaneously. Trees were the cradle of early civilisation, providing people with the materials to build shelters, make tools and generate heat. Bands of hunter-gatherers made good use of Wales' rich natural resources, and were able to build settlements high in the hills thanks to a warm period that extended into the middle Bronze Age. Remains of Bronze Age culture, including settlements and burial chambers, can be found right across Wales, including at many of the sites featured in this guide, such as Coed Cors-y-gedol [63], Coed Nercwys [84] and Ceri Forest [157]. The remarkable cromlech at Parc le Breos Cwm and Coed y Parc [259] and the dolmen at Ty Canol [211] are among the best preserved prehistoric features in Wales. Historians think that the cooling period that occurred around 3,000 years ago, with wetter summers and lower temperatures, led to hardship and social unrest, and the first hilltop forts were constructed by rival tribes. Meanwhile, the Celtic peoples had firmly settled in Wales, with the birth of the early Welsh language happening at about the same time (700 BC).

Five years after invading Britain, the Romans turned their full attention to the 'troublesome' Celts, who had made Wales their stronghold, beginning with an invasion into north-east Wales in AD 48. Hill forts, overnight marching camps and the Romans' famously straight roads were constructed, and their remains are still visible today at many locations, including Roman Camp [11], Banc-y-Castell at Bwlch Nant yr Arian Forest [112], Cae Gaer at Esgair Ychion [158], Trawsgoed Fort at Black Covert [121],

Burfa Bank [171], and the largest fort at Coed Craflwyn and Dinas Emrys [41]. The all-conquering Roman army completed its invasion with the defeat of the Celts in AD 79, yet this only marked the beginning of a long period of territorial disputes between the Celts (Welsh) and the Angles (English). The Romans discovered valuable minerals in Wales, such as gold at Dolaucothi [208], so for centuries to come there was good reason for other powers to seek a share in this wealth.

Some 700 years later (AD 757–796), the Anglo-Saxon king of Mercia is thought to have commissioned the extraordinary linear earthwork known as Offa's Dyke to define the boundary between the Mercia territory and the Welsh kingdom of Powys. It still follows the modern Wales–England border quite closely. As a modern long-distance footpath, it also provides a wonderful route through the countryside, taking in many forest sites, including Castell y Waun (Chirk Castle) [98], Clwyd Forest [83], Coed Llandegla [89], Coed Llangwyfan [80], Llanymynech Rocks [137] and Trevor Hall Wood [94]. A lesser-known earthwork, Wat's Dyke, passes through Knolton Wood [97].

The Normans were next to want a share in the wealth of Wales, invading in 1067–1081, with multiple movements continuing for the next two centuries. Castles were constructed at strategically important places, often with commanding views over rivers and roads, making use of natural features such as steep terrain. Forests continued to provide crucially important materials for their construction, and were vital as a source of food and for hunting by wealthy Normans, and a staple product for baking and heating. The often steep and inaccessible land surrounding these castles later helped protect trees from development or livestock grazing, such as at Cilgerran

An early twentieth-century abandoned mine dram (cart) at Dolaucothi [208].

Castle and Forest Wood [205], Penrice Castle near Mill Wood [260] and Pennard Castle above Pennard Cliff and Northill Woods [261].

The bark of one of the dominant tree species across much of Wales, sessile oak, is rich with tannic acid (or 'tannins'). It provided a major ingredient for tanning leather, an essential material for early industry (e.g. harnesses for working horses) and personal effects (e.g. belts and armour). The process involved felling oak trees and stripping their bark, which was stored for several months until dry. It was then ground to a rough powder and added to vats of water, where the raw hides were soaked in progressively strong tanning solutions in a process that took about 12 months.

Beginning with the Romans, the tanning industry grew to be significant, with oak managed specifically for the tanning industry, effectively as coppice coupes (small areas of woodland coppiced at the same time), creating the first industrial-scale impact on Welsh woods.

With the advent of the Industrial Revolution and Britain's insatiable demand for timber to feed its war machine as a global superpower in the seventeenth and eighteenth centuries, the country's remaining woodlands were critically important for fuel, gunpowder manufacturing and construction. The natural resources of Wales were also central to this industrial growth, especially ironworks and coal mining in the south

The Felin Fawr waterwheel near Coed Maenarthur [124] once provided power for a local lead mine.

of the country. The previously rural society changed rapidly, and people migrated from across Wales to work in the industrial heartlands of Glamorgan and Monmouthshire. See Green to black and back (pp.230–31).

The First World War was a turning point for forestry and the upland landscape in Wales. The shortage of timber became a national matter of strategic importance, caused by unprecedented demand and blockades of global imports from Britain's overseas territories. This led ultimately to the creation of the Forestry Commission (FC) in 1919, and a huge effort to create more productive forests, especially in the uplands of Wales, where the depressed state of agriculture led to attractive land prices, aiding purchase by government. Innovations in silviculture and tree breeding resulted in extensive plantations of exotic tree species such as Sitka spruce and lodgepole pine in the uplands, and larch and Douglas fir in the lowlands. One of the first plantations created by the FC was in 1921 at Gwydir Forest [34] in North-West Wales, while grants were made available to private landowners to encourage afforestation at scale. Much of the softwood produced in these new plantations was destined to provide pit-props for the coal industry. At the time, there was little appreciation for the historic environment or for wildlife. Ancient woodland was replaced with productive plantations (see pp.188–89), landscapes covered with ugly and impenetrable monocultures, watercourses ruined by acidification, and historic sites buried under dense forest.

Today, forestry in Wales is cutting edge in every sense. Modern forestry practice is incomparable to the past, and the 'art and science' of forestry balances the need to produce fibre and structural timber, while improving our own health and well-being, beautifying our landscapes, cleaning our air and water, and repairing and maintaining habitats for wildlife. The public forest estate in Wales was transferred from the Forestry Commission to Natural Resources Wales in 2013.

Forestry and silviculture

A guide to forests would be incomplete without the mention of silviculture or, more broadly, forest management, practised by foresters. Forestry is often described as both a science and an art. Deep knowledge of tree biology, soils and geology, and of the wildlife that depends on the ecosystems in a forest, are all required by the modern forester, plus an appreciation of social health and well-being, and an ability to calculate embodied carbon while projecting the future impacts of climate change. Add this to an awareness of landscape design and the art of anticipating the needs of future human society, and it is easy to see that being a forestry professional is a deeply satisfying and life-changing career in every sense.

The days of planting and managing forests purely for profit are of a bygone age, though reputations can be hard to turn around from the era of wall-to-wall afforestation with exotic conifers, especially while some of the legacies from those twentieth-century practices linger on in Welsh landscapes. What has changed fundamentally is the role of the forester and the business of forestry. Britain has led the way globally in forest certification, especially in the development of the independently audited UK Woodland Assurance Standard, through schemes like those run by the Programme for the Endorsement of Forest Certification (PEFC) and the Forest Stewardship Council (FSC). According to official statistics provided by government, three-quarters of the volume of all timber felled in Wales is certified.

Professional standards are vigorously upheld by the Institute of Chartered Foresters, and becoming a chartered professional means as much as it does in any career. While the ghosts of the past may haunt foresters still (or the mention of lumberjacks and lumberjills in plaid shirts may draw the odd nudge and wink), it is our foresters we must turn to if we are to revive Britain's forests, and not only find space for

Timber stack.

nature during the unprecedented challenge of the climate crisis, but also help save us from ourselves. We have only recently begun to appreciate the roles that forestry can play in alleviating flooding, cleaning our air and purifying our water. Forests lock up carbon in their soil and trees, and in the timber we grow and use as a substitute for harmful humanmade materials.

Forestry is worth almost £500 million to Wales' economy, and approximately 10,000 people are employed in all branches of forestry, as forest managers and wildlife rangers, in tourism and recreation roles, and of course in the wood chain. Twelve sawmill sites operate in Wales, plus there are many more mobile sawmills operating across the country. Some 668,000 tonnes of softwoods are felled annually in Wales from the public forest estate, and 581,000 tonnes from privately owned sites.

Welsh language and place names

While this book is written in English, the Welsh language, Cymraeg, deserves to be featured and promoted. Where the names differ, this guide provides names for towns as Cymraeg/English, as both may be seen on maps and road signs. The names of afonydd (rivers) are given in Cymraeg, followed by their English name where different (a full list can be found in the Appendices, p.270).

The Welsh language is most commonly spoken in Gwynedd (64 per cent of the population speak Cymraeg), and quite common in Conwy, Denbighshire, Ynys Môn/ Isle of Anglesey, Carmarthenshire, north Pembrokeshire, Ceredigion, parts of Glamorgan and some areas in Powys. Cymraeg is a mandatory curriculum subject in all schools in Wales.

When in Cymru (Wales) – pronounced 'Kum-ree' – visitors might benefit from learning to speak or understand a little Cymraeg (Welsh) – pronounced 'Kum-raig'. It is mostly a phonetic language, and all letters are usually pronounced. The challenge for non-Welsh speakers is learning how the letter combinations should sound! The emphasis for pronunciation is often placed on the penultimate syllable, for example 'Ban-eye Brih-chein-eeog' for the Bannau Brycheiniog (Brecon Beacons). The Welsh for wood, *coed*, is pronounced 'coyd'. Have a go, and your efforts will be hugely appreciated by Cymraeg speakers.

Some geographical vocabulary is very helpful in bringing place names to life, as many features and sites have a literal meaning. For example, Aberystwyth means the confluence of the Afon (river) Ystwyth, while Coed y Cwm [109] means wood of the valley. A list of helpful Cymraeg terms is provided in the Appendices (pp.269–271).

Forest ownership

Britain has an unusually high degree of private forest ownership, with almost 75 per cent of its forest area owned by private individuals, companies and non-governmental organisations. In Wales, the public forest estate is managed by Natural Resources Wales, while other significant woodland-owning institutions include the National Trust, Ministry of Defence, Crown Estate, Welsh Water, Network Rail, the RSPB, The Woodland Trust and the Church Commissioners for England.

Natural Resources Wales (NRW)

NRW (Cyfoeth Naturiol Cymru) is a Welsh Government-sponsored body set up in 2013 when Forestry Commission Wales was merged with other environmental bodies. It has responsibility to care for the natural resources of the country, including the public forest estate, which extends across 309,000ha (47 per cent of Wales' forest cover). There are 102 sites managed by NRW in this guide. The Welsh Government has committed to creating the National Forest for Wales, which involves creating new woodland and restoring ancient woodlands, with the aim to form a network of woodlands throughout Wales. This network will bring social, economic and environmental benefits by creating woodlands at scale and by forming corridors for wildlife to move more freely through the landscape. Fourteen sites are currently included:

Gwydir Forest [34]
Clocaenog Forest [91]
Coed y Brenin Forest Park [59]
Dyfnant Forest [139]
Dyfi Forest [70]
Bwlch Nant yr Arian Forest [112]
Hafren Forest [156]
Coed y Bont [126]
Presteigne forests, including Radnor Forest [162]
Brechfa Forest Garden [213]
Afan Forest Park [254]
Spirit of Llynfi Woodland [274]
Wentwood [317]
Wye Valley Woodland [multiple sites]

NRW is active in encouraging access and recreation to many of its sites, providing car parks, waymarked trails and interpretation boards.

Ministry of Defence (MOD)

The Defence Estate manages land on behalf of the MOD, with nine sites across Wales. See Crychan Forest [206] and Halfway Forest [187].

National Trust (Cymru)

The charity owns some 50 properties across Wales, including historic houses, coastline, parks and gardens. There are 24 forest sites owned by the National Trust featured in this guide, including Aber Mawr [217] and Stackpole [257] on the coast; Bodnant Garden [12], famous for its large trees; Cwm Sere [189], which harbours rare endemic whitebeams; and Pennard Cliff and Northill Woods [261] below the dramatic ruins of Pennard Castle.

Wildlife Trusts

There are five Wildlife Trusts in Wales, each operating as an independent charity, with 85 sites featured in this guide covering 928ha of forest: Gwent (14 sites/204ha), Montgomeryshire (5 sites/43ha), North Wales (18 sites/172ha), Radnorshire (17 sites/56ha) and South and West Wales (31 sites/453ha).

The Woodland Trust (Coed Cadw)

The charity owns more than 100 sites across Wales, covering about 2,900ha. There are 25 sites owned by The Woodland Trust featured in this guide, the largest being the precious temperate rainforest at Coed Felenrhyd and Llennyrch [48] and a significant area of Wentwood Forest [317], which it manages in partnership with NRW.

RSPB (Cymru)

There are seven forest sites owned by the RSPB featured in this guide, the largest being more than 1,000ha at Lake Vyrnwy [138], with others at Arthog Bog [68], Carngafallt [167], Coed Garth Gell [65], Cwm Clydach [246], Gwenffrwd-Dinas [201] and Ynys-hir [103].

Community ownership

There are about 138 active community woodlands across Wales, mostly on land owned by local authorities. A community-led group takes an active role in the management of a woodland, which it may own, lease or work in, under the owner's permission. The organisation Llais y Goedwig provides a central coordinating role for community woodlands in Wales. The highest number of community woodlands are in South Wales in Blaenau Gwent (15), Bridgend (11) and Pembrokeshire (10), while in the North-West region there 11 groups in Gwynedd.

The Afon Tywi/River Towy flows through Tywi Forest [127].

HOW TO USE THIS GUIDE

In this guide, Wales is divided into seven regions. Six of these are widely adopted by the Welsh Government, while the usual 'Central Wales' is here split into two regions: Central-West and Central-East. Within each region, sites are ordered from west to east, then north to south, like reading a book. Each site is allocated a number, and their geographical location is marked on the regional maps on pages 276–82. In the main text, these site references have been noted within square brackets [x]. Each site entry begins with key information summarising useful facts and figures:

- **THE NEAREST CITY/TOWN/VILLAGE** is provided to help with locating the site on a map or when searching with SatNav. Names are provided in Cymraeg/English. The county of the site is also included, but note that the nearest town may not fall within that county.

- **MAP REFERENCE** ('Map ref') is the number given to each of the 325 sites to help identify them on the maps provided.
- **SYMBOLS** to indicate features of special or notable interest.

nature   scenic   history

- **OWNERSHIP:** either public (detailed) or private (detailed where possible). Note that ownership by charities and community groups is deemed private, even though they are generally owned for public good.
- **DESIGNATIONS:** formal designations for archaeology, landscape or wildlife (for further details see Site Designations, p.268). Note that descriptive terms, particularly 'nature reserve', can be added by site owners, but these are not necessarily formal designations in the sense applied here. Where none are listed, the site has no formal designations.

Craig y Cilau [193].

HPG: Historic Park and Garden
IPA: Important Plant Area
LNR: Local Nature Reserve
NL: National Landscapes (formerly AONBs)
NNR: National Nature Reserve
NP: National Park
SAC: Special Area of Conservation
SM: Scheduled Monument
SPA: Special Protection Area
SSSI: Site of Special Scientific Interest

- **AREA:** size of forest or tree cover in hectares (ha). Multiply by 2.47 for acres.
- **FOREST TYPE:** the broad category of woodland; mainly broadleaved, conifer or mixed (at least 30 per cent of one or the other). In addition, a special category is used for temperate rainforest sites.
- **FOREST LOCATION:** the central point of the forest site. Note: this is not necessarily a precise location for navigation.

  Forest location: Ordnance Survey six-figure grid reference.
  Explorer Map: OS Explorer Map (1:25,000 scale) number.

- **EASE OF ACCESS:** this grading is used to indicate relative difficulty of access within the woodland and from the nearest access point such as a car park.

  Easy sites are typically a short distance from the access point and can be enjoyed without specialist walking gear. The site may be acceptable for wheelchairs or for families with buggies. See Access for disabled people, p.26.
  Moderate means the distance to/within the site will be further than those marked EASY. Stout walking shoes and waterproofs may be a wise precaution. Trails will not always be waymarked. See Keeping Safe, p.27.
  Difficult sites are likely to be remote, requiring considerable effort to reach and probably strenuous to explore. Proper walking equipment recommended. Map-reading skills required.

- **ACCESS POINT:** where to park a vehicle or gain access to a forest site. Note: this may be some distance from the forest site.

  Grid ref – pinpoint location based on Ordnance Survey ten-figure grid reference.
  what3words – unique three-word code referencing the Access point (3m accuracy). Works best outdoors via its mobile app, or in car SatNav when available.
  Postcode – use with a SatNav to navigate to the vicinity of an access point. Note: postcodes often have limited use in rural areas – **do not rely on postcodes alone!**

Alongside this book, please visit The Forest Guide online:

gabrielhemery.com/forest-guide

Readers can enjoy free access to an interactive online map to all sites featured in this guide. Use the password *copsewoodforest* to gain full access.

This guidebook is the second in a series featuring the copses, woods and forests of Britain published by Bloomsbury Wildlife. *The Forest Guide: Scotland* was published in 2023, and *The Forest Guide: England* is forthcoming.

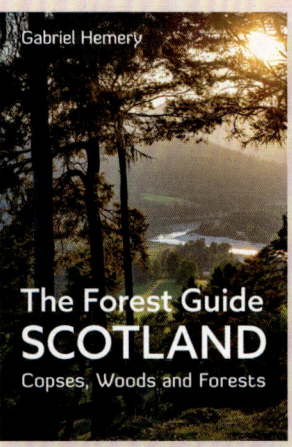

How to Use This Guide

The view from The Kymin above Beaulieu Wood [297]

ACCESS TO FORESTS

Take nothing but pictures – Leave nothing but footprints – Kill nothing but time

Unlike in Scotland, the laws governing public access to land in Wales or England do not provide for a 'right to roam'. However, one-fifth of Wales is designated open access land under the Countryside and Rights of Way Act (CROW), and these areas are marked with orange-shaded borders on OS maps. Forest sites within CROW open access land are usually part of the public forest estate, managed by National Resources Wales, or occasionally registered common ground. Otherwise, visitors to privately owned woodlands must rely either on public rights of way or permitted access. Public rights of way include public footpaths, bridleways or byways, and these are clearly marked on OS Explorer (1:25,000) or Landranger (1:50,000) maps.

Landowners may choose to provide permitted access, which also means that they can decide to withdraw it without notice. Such forms of permitted access may be provided under open terms for any visitor, or restricted for certain groups of people (e.g. local people, sports clubs), specific purposes (e.g. shooting, biological monitoring) or limited to certain times of the year. Any such permitted access is not indicated on maps, and the visitor should check either with a landowner's website (if available) or local notices.

All sites included in this guide are publicly accessible unless expressly described in the text, but in many cases only by using public rights of way. Please observe any local signs and use a map if you are unsure which areas you can access.

If you are cycling or horse riding, keep to bridleways to comply with the law and to minimise any damage. If you have a dog with you, keep it on a short lead or close at heel during the spring (April to July) so that breeding birds are not disturbed. Livestock

A walker passing through giant conifers at Hafod [120].

might be present in some forests and woods, so take care if you come across any animals. Be careful not to trample young trees. Access to forests may be legally barred by landowners during forestry operations.

Forest health

Tree pests and pathogens are becoming ever more a serious problem in Britain. Some of the blame falls squarely on our reliance on importing young plants from overseas, or even in global trade itself, with unwanted creatures hitching rides in packing crates and shipping containers. The climate crisis is playing a role too, in that the changing conditions can favour pests and pathogens that before were unable to thrive.

Follow these steps to conserve forest health:

- before and after every visit to a woodland, clean mud and leaves from your footwear, or from pushchairs, bikes, cars, dogs and horses;
- do not remove leaves, live plants or wood.

Dogs

Dogs are often permitted at most sites, unless in the grounds of a house or venue such as a botanic garden. If visiting a tourist destination, dog owners are advised to check details on any website for the venue. Dogs are more usually banned from nature reserves, but sometimes permitted on a lead. Elsewhere, while dogs might be permitted, owners should be very aware of the impact that loose dogs can have on wildlife, particularly ground-nesting birds during the breeding season. Signs are often erected by owners or managers at vulnerable places to politely request that a dog is always kept on a lead. Such requests should always be heeded, even if you think your dog is well behaved. Dog waste must be removed and disposed of safely.

Camping

Opportunities for wild camping are limited in Wales, with specific landowner permission required within the national parks.

Always make sure you:

- remove all litter;
- leave no trace of your stay;
- avoid lighting a fire or using a stove in or near a woodland;
- bury your poo and don't use soap near watercourses.

Military

The military undertake live firing exercises across nine areas in Wales, and these are clearly marked on Ordnance Survey maps. Always take note of advice from range staff, troops and warning signs. If in doubt, look for an alternative route or turn back. Red flags (in daytime) and red lamps (at night) indicate live firing areas, which might not be fenced. Do not enter a range if flags are raised or lamps lit. Be careful when crossing the land as there could be trenches or voids, and never pick up objects in these areas as they could be harmful.

Access for disabled people

If you use a motorised vehicle, take special care not to disturb wildlife, and remember to follow the Highway Code. Natural Resources Wales provides online resources promoting accessibility, including adopting a trail grading system, which is implemented on the ground using signage. The Phototrails website provided by the Fieldfare Trust is another useful resource.

Landowner responsibilities

Landowners are required to allow access to public rights of way, and are not permitted to obstruct entry or routes across their land, nor to intimidate members of the public.

The Precipice Walk near Nannau [64].

A treescape in Ceredigion.

KEEPING SAFE

Exploring the outdoors

Most of the forest sites in this guide are relatively easy to access, but some are in remote locations. Those marked as Difficult are in hard-to-reach places, which require a good level of fitness, proper personal equipment and an ability to navigate using map and compass. You should know what to do if you get into trouble, and not rely solely on a mobile phone. It is worth registering for a special text service, which allows you to text '999' for help, because a text message can sometimes be sent when a call cannot be made. You must first register to use this service before you venture out (i.e. not at the moment you need it!). Simply text the word 'register' to 999. Follow the instructions that you receive in a subsequent text message to complete registration.

Middle right: *Ixodes ricinus*, more commonly known as the castor bean, sheep or deer tick.

Ticks and Lyme disease

It's important to be aware that ticks in the countryside sometimes carry an infectious disease called Lyme disease. To reduce the likelihood of ticks biting, wear long trousers tucked into your socks, or spray insect repellent on exposed areas. Try to avoid wading through long grass or bracken. If you find a tick, remove it as soon as possible. Familiarise yourself with the symptoms of Lyme disease and follow the advice on the NHS website.

The North-West, encompassing Ynys Môn/Isle of Anglesey, Conwy, and Gwynedd is the third largest, yet – with a little under 11 per cent forest cover – the least-wooded region in Wales. What it lacks in quantity, it certainly makes up in quality, with the greatest number of temperate rainforests for any region. These include sites such as Coed Felenrhyd and Llennyrch [48], whose sessile oak trees are festooned with mosses, lichens, hornworts and ferns, and home to birds that are rare elsewhere in Britain. More than half the region falls under the national park for the

The Precipice Walk and the Afon Mawddach near Nannau [64].

Eryri (Snowdon) mountains. The region includes several very extensive conifer plantations, including the first created in Wales by the Forestry Commission, Gwydir Forest [34] planted in 1921, and Dyfi Forest [70], which at more than 6,000ha is the second largest in this guide. The forests of the North-West are also one of the last bastions for our native red squirrel, now classed as an endangered mammal, although visitors have a good chance of seeing one in Ynys Môn at Coed Cyrnol [17], Lawr yr Afon [1], Llyn Parc Mawr [23], Newborough Forest [26] and Plas Newydd [20].

NORTH -WEST

YNYS MÔN/ISLE OF ANGLESEY, CONWY, GWYNEDD

SITES 1–72

LAWR YR AFON

CEMAES, YNYS MÔN/ISLE OF ANGLESEY

MAP REF 1

Ownership: Public: Llanbadrig Community Council **Designations:** **Area:** 2ha	**Forest type:** broadleaved **Forest location:** SH371932 **Explorer Map:** 262 **Ease of access:** Easy	**Access point:** SH3731193525 ///points.trinkets.vaccines LL67 0NG

Running through the centre of the village, Lawr yr Afon (which translates as 'down the river') provides precious habitat for nature and hugely valued green space for local people. The small wood of native trees along the banks of the Afon Wygyr, the most northerly river in this guide, is managed by volunteers for Cemaes Bay Nature for All. Common woodland birds and red squirrels can be seen from the broad paths, which are suitable for buggies and wheelchairs, and enjoyed from the many resting places. The remains of the old brickworks' narrow-gauge railway can still be seen in places, while its chimney can be reached at the top of the site via a tunnel underneath the A5025. The Access point is a private car park (charged).

TRAETH YNYS Y FYDLYN

LLANFAIRYNGHORNWY, YNYS MÔN/ISLE OF ANGLESEY MAP REF 2

Ownership: Private Designations: NL Area: 12ha Forest type: conifer	Forest location: SH297918 Explorer Map: 262 Ease of access: Moderate	Access point: SH3032191451 ///gently.centuries.promoting LL65 4LT

The forest growing near the picturesque beach of Ynys y Fydlyn (owned by the National Trust) is the most remote on Anglesey. The plantation of Corsican pine surrounds Llyn y Fydlyn, a freshwater lake immediately behind the beach, with bright yellow gorse scrub growing along its fringes. On higher ground, where relentless sea winds rake the trees, the pines grow as krummholz (stunted and deformed trees). From the car park (Access point), walk through the kissing gate and head towards an old pillbox, following the Wales Coast Path. On a clear day, the Isle of Man is visible to the north. The lighthouse on the Skerries comes into view lower down the path. Spend a moment resting on the beach and watch ravens wheeling overhead.

Traeth Ynys y Fydlyn [2].

BRYN PYDEW

BRYN PYDEW, CONWY   MAP REF 3

Ownership: Private: North Wales Wildlife Trust
Designations: SAC, SSSI
Area: 14ha

Forest type: mixed
Forest location: SH816798
Explorer Map: OL17
Ease of access: Easy/Moderate

Access point: SH8188279807
///opera.lonely.tend
LL31 9QA

Stinking hellebore at Bryn Pydew [3].

This is a beautiful small woodland dominated by ash and yew with occasional holly, situated on limestone pavement surrounded by a golden fringe of gorse when in flower. Clearings in the woodland are managed for the extraordinary stinking hellebore, which grows among the gorse and juniper. Grykes in an area of limestone pavement shelter hart's-tongue fern and invertebrates, and together with the grassland the site harbours more than 600 moth and 22 butterfly species, and hosts a strong population of glow-worms. Six species of orchid are known to grow here, including green-winged, and many other rare plants including autumn ladies tresses and dark-red helleborine (a plant in the orchid family). The Access point is a lay-by immediately next to the site.

BODLONDEB WOODS

CONWY, CONWY   MAP REF 6

Ownership: Public: Conwy County Borough Council
Designations: LNR
Area: 7ha

Forest type: broadleaved
Forest location: SH778781
Explorer Map: OL17
Ease of access: Easy

Access point: SH7790777882
///issued.reclined.billiard
LL30 9SB

Overlooking the southern banks of the Conwy estuary, the woods at Bodlondeb were once a part-private estate but are now managed as an LNR by the borough council. There is an interesting range of tree species in the wood, including evergreen holm oak and an avenue of holly trees among the ash, beech, lime, oak, sycamore and wild cherry. There are specimens of Scots pine and yew among the broadleaves. There is a good diversity of wildlife to see and hear, including tawny owl at dusk. The wood is easily reached from town, while there is ample parking (Access point) nearby. A network of easy-to-follow paths pass through the wood, although there are some moderately steep sections, including with steps. The North Wales Path passes along the wood's boundary with the river, providing a gentle alternative. There are attractive views across the estuary to Deganwy and the Great Orme. Look for heron and cormorant resting in the trees or feeding on the estuary.

PENRHOS COASTAL PARK

CAERGYBI/HOLYHEAD, YNYS MÔN/ISLE OF ANGLESEY   MAP REF 4

Ownership: Private **Designations:** NL **Area:** 34ha **Forest type:** mixed	**Forest location:** SH274810 **Explorer Map:** 262 **Ease of access:** Easy	**Access point:** SH2751680617 ///vehicle.lobby.suave LL65 2JE

The mixed woodland at Penrhos, planted in 1816, sits alongside freshwater lakes, coastal mudflats at Beddmanarch, sandy beaches, rocky promontories and open sea. The woodland includes a wide range of species, including ash, alder, beech, oak and sycamore, with some grand fir and Sitka spruce. Ramsons and bluebells carpet the woodland in spring, and a diverse range of fungi can be discovered in autumn. It is a popular site with birdwatchers, with rarities often sighted. Keep an eye on the tree canopies for red squirrels. Despite being named a nature reserve in the 1960s and lying within a designated NL, a significant proportion has recently been under threat from development. The site is much loved by local people and a popular stopover for those using the Caergybi/Holyhead to Dublin ferry. National Cycle Route 8 passes through the west of the wood.

Penrhos Coastal Park [4].

PWLLYCROCHAN WOODS

BAE COLWYN/COLWYN BAY, CONWY   MAP REF 5

Ownership: Public: Conwy County Borough Council Designations: LNR Area: 21ha	Forest type: broadleaved Forest location: SH843785 Explorer Map: OL17 Ease of access: Easy/Moderate	Access point: SH8427078656 ///nests.paint.cattle LL29 7HE

Surrounded on all sides by housing at the southern side of Bae Colwyn/Colwyn Bay, this dingle is the remains of ancient woodland, once owned by the Pwllycrochan Estate but now owned and managed by the local council as an LNR. Most of the woodland is broadleaved, with many different species including mature ash, oak and sweet chestnut, and occasional Scots pine. Many of the oaks are tall and branch-free, presumably pruned for timber production long ago while under previous ownership. This is a good place for a fungal foray, especially in autumn, and to spot common woodland birds throughout the year. To the west of the wood is a tumbling stream and pool, which is the origin of the wood's name, for *pwll* is Welsh for pool, while *crochan* translates as cauldron (the pool must have once been less tranquil!). From the top of the wood, the sea and offshore wind turbines can be glimpsed beyond the town. There are multiple access points to the wood, with the central section having interpretation signs. Informal car parking is available near the Access point in Pwllycrochan Avenue, where the extraordinary sight of a large specimen of a strawberry tree growing from the middle of the road is a surprising delight.

The strawberry tree near the entrance to Pwllycrochan Woods [5].

PENTRAETH FOREST

PENTRAETH, YNYS MÔN/ISLE OF ANGLESEY   MAP REF 7

Ownership: Private, and Public: Natural Resources Wales Designations: NL Area: 250ha	Forest type: mixed Forest location: SH544789 Explorer Map: 263 Ease of access: Moderate	Access point: SH5337378399 ///cherish.measuring.private LL75 8YH

Most of this mixed forest is privately owned, with only the eastern end being part of the public forest estate. A network of forest rides and footpaths criss-cross the site, one of which climbs close to the top of Mynydd Llwydiarth (158m), offering fine views north to Red Wharf Bay and the glittering sea. An undesignated ancient hill fort lies just south-west of the top but is hidden among conifers. Whichever route is followed, walk quietly and look above to the tree canopies in hope of spotting a red squirrel. There is room for a single car to park (Access point) next to the entrance at the western end of the forest.

BODFFORDD

BODFFORDD, YNYS MÔN/ISLE OF ANGLESEY   MAP REF 8

Ownership: Public: Natural Resources Wales Designations: Area: 68ha	Forest type: mixed Forest location: SH452782 Explorer Map: 263 Ease of access: Easy	Access point: SH4522978212 ///vegetable.hinders.boomer LL77 7RQ

Enjoy a gentle stroll around Llyn Cefni, the second-largest humanmade lake on Anglesey, built to store drinking water for the island. It is bisected by a disused railway line. Its shores were until recently surrounded by mature forest plantations, but the arrival of the deadly pathogen *Phytophthora ramorum* has resulted in the felling of substantial areas where larch trees were infected. The stands are being replaced with mixed broadleaves and conifers, which will create a beautiful ring of trees around the water in future, and new habitat for the red squirrel.

COED BRYNDANSI

DOLWEN, CONWY   MAP REF 9

Ownership: Private Designations: SSSI Area: 92ha Forest type: conifer	Forest location: SH855740 Explorer Map: OL17 Ease of access: Easy/Moderate/Difficult	Access point: SH8549973852 ///cove.sues.abolish LL29 6AU

Follow surfaced forest tracks as they contour around a small hill in a little-visited area of Welsh countryside. This is a place to enjoy peace and solitude just 5km south (as the crow flies) of the holiday crowds of Bae Colwyn/Colwyn Bay. Hidden away in the north-west corner of the forest is a low-lying peaty mire and SSSI at Llyn y Fawnog.

BODNANT GARDEN

LLANSANFFRAID GLAN CONWY/GLAN CONWY, CONWY   MAP REF 12

Ownership: Private: National Trust Designations: Area: 20ha	Forest type: mixed Forest location: SH798720 Explorer Map: OL17 Ease of access: Easy	Access point: SH8020972327 ///sloping.rules.altering LL28 5RF

In 1887, Victorian Henry Pochin established an arboretum in the sheltered gardens on the east bank of the River Conwy. Collections of broadleaves, including acers (maples), magnolias and rhododendrons thrived, as did conifers introduced from North America. Today, the pinetum in Yew Dell includes several champion trees including a 50m-tall giant redwood. The tallest tree in Wales (a 51m-tall coast redwood) once grew here but was toppled by Storm Arwen in 2021. Entry charge.

Foliage of giant redwood.

BANGOR FOREST GARDEN

ABERGWYNGREGYN, GWYNEDD

MAP REF 10

Ownership: Private: Bangor University **Designations:**	**Area:** 8ha **Forest type:** broadleaved **Forest location:** SH654732	**Explorer Map:** OL17 **Ease of access:** Easy **Access point:** See details below.

Forest gardening is a technique used in permaculture, which in turn is a philosophy based on caring for people, nurturing the Earth and fair sharing. It combines silviculture (forest management) with agriculture, in a management technique known as agroforestry. Its aims are to produce food, fibre and fuel sustainably while promoting biodiversity. Owned by the university, Bangor Forest Garden is run by a voluntary group that meets on the second Sunday and last Wednesday of every month. Visitors are welcome to attend any of these sessions as a volunteer, but otherwise the site is not open to casual visitors. Details can be found on social media. The site is positioned between the sands of Traeth Lafan/Lavan Sands (LNR, SAC and SSSI) to the north, and the Eryri/Snowdonia National Park immediately to the south.

ROMAN CAMP

GARTH, GWYNEDD

MAP REF 11

Ownership: Public: Gwynedd Council
Designations: SM, SSSI
Area: 8ha
Forest type: broadleaved
Forest location: SH581728
Explorer Map: OL17
Ease of access: Easy/Moderate
Access point: SH5835572994
///august.polar.recorder
LL57 2SR

This small wooded hill near Garth is popular with local people, who know it as Roman Camp. The site is designated as a Scheduled Monument but has never been excavated. It earnt its name after a Roman coin was discovered here in the 1960s, but it could well be an Iron Age hill fort or the site of a medieval castle. From the top of the hill there are extensive views, looking south over Bangor and beyond to Penrhyn Castle, Bangor Mountain, and the mountains of the Eryri/Snowdonia National Park. In the opposite direction lie the trees of Nantporth [14] above the shoreline of the Afon Menai/Menai Strait and beyond the two bridges linking to Ynys Môn/Isle of Anglesey.

The view of Bangor and Penrhyn Castle from Roman Camp [11].

COED CRAFLWYN

ROWEN, CONWY — MAP REF 13

Ownership: Private Designations: Area: 6ha Forest type: mixed	Forest location: SH764722 Explorer Map: OL17 Ease of access: Easy/Moderate	Access point: SH7654972233 ///unleashed.fearfully.implore LL32 8TP

Large, old coppiced oak stools and occasional wide-spreading holly trees can be found under the tall beech and Scots pine planted here during the 1950s on this once-ancient woodland site. Prominent wood banks and old trackways hint at the long history of this former working coppice woodland. The site lies just outside the boundary of the Eryri/Snowdonia National Park. A clearing at the top of the hill offers a fine view of Tal y Fan (610m), an outlying peak of the Carneddau mountains in North Wales.

NANTPORTH

BANGOR, GWYNEDD — MAP REF 14

Ownership: Private: North Wales Wildlife Trust Designations: SSSI Area: 5ha	Forest type: broadleaved Forest location: SH569723 Explorer Map: OL17 or 263 Ease of access: Moderate	Access point: SH5743672549 ///synthetic.radiates.mint LL57 2BE

The trees of this strip of coastal woodland come right to the shores of the Afon Menai/Menai Strait. The ash-dominated canopy includes occasional oak, plus aspen and birch. It also contains the entire global population of the Menai whitebeam (*Sorbus arvonensis*). Only 30 specimens are thought to exist. Look for them on the steep slopes, where there is plenty of sunlight. Nuthatch and jay frequent the woods, while the calls of redshank and curlew feeding on the mudflats echo through the trees.

Steep and narrow paths include many sections of steps. The south end of the reserve can be reached on foot from Bangor town by following the Wales Coast Path. There is limited space for cars on Gorad Road at the northern end (Access point).

Nantporth [14] seen from Roman Camp [11].

The trees of Coedydd Aber [15].

COEDYDD ABER

ABERGWYNGREGYN, GWYNEDD   MAP REF 15

Ownership: Private: Bangor University, and Public: Natural Resources Wales **Designations:** NNR, NP, SAC, SM, SSSI	**Area:** 346ha **Forest type:** broadleaved **Forest location:** SH666709 **Explorer Map:** OL17 **Ease of access:** Easy/Moderate	**Access point:** SH6624072004 ///smashes.retraced.wanted LL33 0LP

This spectacular woodland grows from near sea level up a valley to 500m altitude, much of it designated as a National Nature Reserve and protected by other designations. The main tree canopy is sessile oak with downy birch and hazel, plus patches of ash and wych elm. The site is important for several rare mosses, including beck pocket-moss and rigid apple moss, plus lungwort lichen. Pockets of carr woodland feature wonderfully old coppiced alders. The many breeding birds recorded in the woodland include pied flycatcher, redstart and wood warbler, while chough are often observed flying on the open moorland. The woodland supports a rich ground flora, including a spectacular show of bluebells in spring. It is an excellent woodland for fungi hunting in the autumn.

At the top of the valley, where ash dominates, water thunders spectacularly from the Carneddau mountains at the Rhaeadr-fawr/Aber Falls (SH6681470045 | ///unlimited.edicts.bearable), which can be reached via a surfaced 4km-return trail.

Prehistoric hut circles and cairns surround the woodland, and nearby the hill fort of Maes y Gaer Camp (SH6632372505 | ///lawyer.ranch.objective) commands spectacular views across the wooded valley and Bae Conwy/Conwy Bay.

The Access point is the lower car park (charge), while another car park suited to disabled parking with facilities is provided in the upper car park (SH6647171903 | ///breathy.squeaking.ribs).

SPINNIES ABEROGWEN

BANGOR, GWYNEDD — MAP REF 16

Ownership: Private: North Wales Wildlife Trust **Designations:** **Area:** 3ha	**Forest type:** broadleaved **Forest location:** SH612721 **Explorer Map:** OL17 **Ease of access:** Easy/Moderate	**Access point:** SH6130272027 ///pairings.eruptions.fiery LL57 3YH

This is a little gem of wet woodland with lagoons next to the tidal mudflats of Traeth Lafan in Conwy Bay. The term 'spinney' means a small woodland, usually adopted for areas of trees created for hunting game. Today, this peaceful site is a haven for birdwatchers, and ideal for spotting kingfisher, heron and little egret, helped by a well-sited bird hide. There is a small space next to the entrance for one or two cars, best kept for those with special access requirements. If possible, park in the car park at the end of the lane (SH6150772389 | ///drill.survived.concluded).

COED DOLGARROG

DOLGARROG, CONWY — MAP REF 21

Ownership: Public: Natural Resources Wales **Designations:** NNR, NP, SSSI **Area:** 69ha	**Forest type:** broadleaved **Forest location:** SH765671 **Explorer Map:** OL17 **Ease of access:** Moderate/Difficult	**Access point:** SH7692567496 ///gifted.chatting.advantage LL32 8QE

Bursting with wildlife yet with a history of human tragedy, this beautiful wood grows on very steep slopes west of the River Conwy. It hosts an unusual array of species for Wales due to its alkaline (lime-rich) soils. The canopy is dominated by beech, with ash, small-leaved lime, sycamore and wych elm, while alder thrives on wetter ground towards the top of the site. Bluebells carpet the woodland floor in spring, while dog's mercury and wild garlic are also common. More unusual plants include broadleaved helleborine and town hall clock. As is often the case in beech woodlands, there is a rich fungi community, notably the rare dewdrop dapperling.

The Access point is a path that begins on the roadside. It follows a giant pair of hydroelectric water pipes running down the hill. After ascending next to the pipes for 300m, the path climbs even more steeply via a series of zigzags. Near the top of the wood, a path heads north to Afon Porthllwyd but do not be tempted to explore the spectacular gorge up close because water is released from Llyn Coedty/Coedty Reservoir without warning, and the resulting flash floods are extremely dangerous. The route can be retraced, or instead climb further up the hill towards the crags,

Coed Dolgarrog [21] seen from the opposite side of the valley from Coed Tan-yr-allt [22].

where peregrine soar, before heading south towards the valley of the Afon Ddu, descending back into the woods where a path heads to the water treatment works at Pont Dolgarrog.

On 2 November 1925, 16 local people lost their lives in the Dolgarrog Dam disaster. A memorial garden (SH7691367697 | ///users.define.dish) includes two short waymarked trails and interpretation boards offering poignant stories, set among the enormous boulders carried down the gorge during the catastrophic flood caused by the breach of two dams.

PLAS NEWYDD

LLANFAIRPWLLGWYNGYLLGOGERYCHWYRNDROBWLL LLANTYSILIOGOGOGOCH/LLANFAIR PG, YNYS MÔN/ISLE OF ANGLESEY

MAP REF 20

Ownership: Private: National Trust	Forest location: SH522702	Access point: SH5174669825
Designations: NL, SM	Explorer Map: OL17	///grades.ballpoint.followers
Area: 49ha	Ease of access: Easy	LL61 6PF
Forest type: mixed		

The beautiful grounds at Plas Newydd include mixed woodlands along the western edge of the Afon Menai/Menai Strait. Red squirrels have a strong foothold here. Across the water lies Vaynol Park [19], and together with the distant peaks of Eryri/Snowdonia the views are very attractive. A prehistoric burial chamber lies within the grounds. An Australasian arboretum features several exotic species of eucalyptus and southern beech. Many of the paths are hard surfaced, while an accessible route leads to a viewing area. Dogs required to be kept on leads. Entry charge.

COED CYRNOL

PORTHAETHWY/MENAI BRIDGE, YNYS MÔN/ISLE OF ANGLESEY

MAP REF 17

Ownership: Public: Menai Bridge Town Council Designations: LNR, NL Area: 4ha	Forest type: mixed Forest location: SH553718 Explorer Map: OL17 Ease of access: Easy/Moderate	Access point: SH5542271914 ///heartburn.exhaling.comb LL59 5EA

Situated on the outskirts of Porthaethwy/Menai Bridge town, this LNR is popular with local people and provides relief to tourists tired after a long drive to Ynys Môn/Isle of Anglesey. The Scots pine trees, accompanied by mixed broadleaves, stretch down to the swirling waters of the Afon Menai/Menai Strait, which retreat to reveal substantial mudflats attracting waterfowl and waders. Look for siskin, goldcrest and even crossbill in the canopies of the pines.

Spotted flycatcher, chiffchaff and blackcap arrive each spring. Red squirrels are present throughout the year. Follow a waymarked path to reach the coast where the Belgian Promenade follows the shoreline, while the causeway can be crossed (keep your eyes peeled for grey seals) to visit the ancient place of worship on Church Island. The views across the water to the mountains of Eryri/Snowdonia are superb. Parking is available at the Access point (charge).

COEDLAN LLANFAIRPWLL

LLANFAIRPWLLGWYNGYLLGOGERYCHWYRNDROBWLL LLANTYSILIOGOGOGOCH/LLANFAIR PG, YNYS MÔN/ISLE OF ANGLESEY

MAP REF 18

Ownership: Private: Anglesey Column Trust Designations: NL, SSSI Area: 1ha	Forest type: broadleaved Forest location: SH534715 Explorer Map: OL17 Ease of access: Easy	Access point: SH5332871542 ///topmost.snuggled.backup LL61 5YL

Not only does the village boast one of the world's longest place names (often shortened to Llanfairpwll), and is the proud birthplace of the British Women's Institute, it also hosts a Napoleonic War memorial. Known as the Marquess of Anglesey's Column, the 27m-tall structure with 115 internal steps is surrounded by ash and sycamore trees. It commemorates Henry Paget, who was second in command under Wellington at the Battle of Waterloo in 1815. The war hero led the allied cavalry, but towards the end of the battle had his leg amputated. When open (charge), the view from the top offers commanding views of the Afon Menai/Menai Strait.

The site is designated an SSSI due to the visible presence of rare blueschist metamorphic rocks.

Sycamore leaves.

VAYNOL PARK

BANGOR, GWYNEDD   MAP REF 19

Ownership: Private: National Trust	Forest location: SH533698	Access point: SH5338869800
Designations: HPG	Explorer Map: OL17	///weary.deadline.final
Area: 40ha	Ease of access: Easy/Moderate	LL57 4FE
Forest type: conifer		

The ancient parkland, also known as Glan Faenol, contains two main blocks of trees at Vaynol Wood and Boathouse Covert. A 3km circular walk, which takes in both, provides ample chance to enjoy the views across the Afon Menai/Menai Strait, including Plas Newydd [20] across the water, while an elevated bird screen and a viewing platform both provide an opportunity to watch waders on the mudflats and even a seal in the open water. The nineteenth-century Gothic-style mausoleum (SH5358670338 | ///melts.blazing.premiums) in the heart of Vaynol Wood is eerily atmospheric and worth a visit. The Access point is not well signposted and reached at the end of a rough-surfaced road.

COED TAN-YR-ALLT

MAENAN, CONWY   MAP REF 22

Ownership: Private: National Trust	Forest type: mixed	Access point: SH7944566527
Designations: SM	Forest location: SH789668	///bordering.asked.defensive
Area: 53ha	Explorer Map: OL17	LL26 0YD
	Ease of access: Moderate	

The commanding views gained over the Conwy Valley from this site have been valued since prehistory. The Iron Age hill fort of Caer Oleu Camp is easily reached from the Access point. Three substantial remains surround two central crags, while the west and east sides are so precipitous that nature provided unrivalled defences to its inhabitants without the need to build any walls. It is well worth continuing through the trees to the spectacular viewpoint at Cadair Ifan Goch, which is reached after a short scramble. From there, the full extent of the steep-wooded sides of the Conwy Valley can be appreciated, including Coed Dolgarrog [21] on the opposite side. The wood contains a wide variety of tree species, especially beech, wild cherry, oak and larch, and their branches are festooned with mosses and lichens. The bluebells here are particularly spectacular in spring.

Lichens thrive at Coed Tan-y-allt [22].

LLYN PARC MAWR

NIWBWRCH/NEWBOROUGH, YNYS MÔN/ISLE OF ANGLESEY

MAP REF 23

Ownership: Public: Natural Resources Wales Designations: Area: 24ha	Forest type: mixed Forest location: SH415674 Explorer Map: 262 Ease of access: Easy	Access point: SH4132666989 ///material.firewall.masts LL61 6WA

Lying at the northern tip of Newborough Forest [26] is an area originally part of the same plantation and now managed for the same owner by the Llyn Parc Mawr Community Woodland Group. It sits just outside the conservation designations of its larger neighbour, but offers a greater chance to observe the native red squirrel, thanks to regular feeding stations. A fine wildlife hide constructed using Welsh-grown Douglas fir looks over a large humanmade lake, providing further opportunities to watch woodland wildlife and waterlife. The woodland is dominated by Corsican pine but the community group is working to diversify tree species by introducing native broadleaves and conifers, and aiming to create more structure in the woodland, with different-aged trees. In time, these changes will help improve the habitat for red squirrels and other wildlife. A circular trail can be followed to reach the hide at the far (north-east) end of the lake.

A robin sings at Llyn Parc Mawr [23].

The wildlife hide at Llyn Parc Mawr [23].

BRAICHMELYN

BETHESDA, GWYNEDD   MAP REF 24

Ownership: Public: Natural Resources Wales Designations: NP Area: 47ha	Forest type: mixed Forest location: SH629658 Explorer Map: OL17 Ease of access: Moderate	Access point: SH6282465389 ///twirls.delays.reaction LL57 3LQ

Braichmelyn Forest was planted for timber production by the Forestry Commission in the mid-twentieth century. Following outbreaks of forest fire and pathogens, the forest is being slowly transitioned from predominantly coniferous to a mixture of species. Lodgepole pine dominates on higher slopes, while Scots pine and Norway spruce are regenerating lower down, together with areas of broadleaves.

Informal access can be enjoyed by following forest rides and public footpaths over the craggy hill. There is room for a few cars at the forest's southern end (Access point). At the northern perimeter nearest the town, hiding among ancient oaks is the giant Bethesda Super Boulder (SH6313266066 | ///medium.earphones.bronze), which offers climbers no fewer than 40 bouldering routes to enjoy.

COED SHED

GROES, CONWY   MAP REF 25

Ownership: Public: Conwy County Borough Council Designations: Area: 8ha	Forest type: broadleaved Forest location: SJ010640 Explorer Map: 264 Ease of access: Easy/Moderate	Access point: SJ0099464261 ///contain.factories.craziest LL16 5RS

Following an unnamed tributary of the Afon Ystrad, this attractive woodland contains many mature oak trees and lights up with flowers in spring. Listen for the whistle of nuthatch and for the two-tone call of chiffchaff. After a short climb down to the valley bottom, a level bridleway can be followed through the wood. A small lay-by has room for one or two cars near a kissing gate into the wood (Access point). The wood could be included in a circular walk from the nearby village of Groes.

NEWBOROUGH FOREST

NIWBWRCH/NEWBOROUGH, YNYS MÔN/ISLE OF ANGLESEY   MAP REF 26

Ownership: Public: Natural Resources Wales Designations: IPA, NL, SAC, SM, SSSI	Area: 654ha Forest type: conifer Forest location: SH404646 Explorer Map: 263	Ease of access: Easy/Moderate Access point: SH4056463443 ///version.noisy.entrusted LL61 6SA

Next to one of the largest dune systems in Britain, whose dunes, slacks (wet hollows) and salt marshes are home to precious rare wildlife, is the large plantation forest of Niwbwrch/Newborough. Planting of its Corsican pine trees began in 1947 to help stabilise the shifting sand dunes and to provide timber. Keep an eye out for red squirrels in their canopies. A selection of waymarked trails through the large forest make navigation relatively easy, although the length of some routes can be more demanding than others. Just north of the main car park (charge) are the remains of the medieval farmstead of Hendai (SH4049863740 | ///wreck.thing.chats).

LLYN CRAFNANT

TREFRIW, CONWY   MAP REF 27

Ownership: Public: Natural Resources Wales	**Forest type:** conifer	**Access point:** SH7559261835 ///track.duty.succumbs
Designations: NP, SM	**Forest location:** SH755618	LL27 0JZ
Area: 170ha	**Explorer Map:** OL17	
	Ease of access: Easy/Moderate	

Not to be confused with the nature reserve Coed Crafnant [56], which also lies with the Eryri/Snowdonia National Park, but far away to the south-west in Gwynedd, this site is part of the Gwydir Forest. Its predominantly coniferous plantations surround the picturesque humanmade lake of Llyn Crafnant. It is a short walk up the lane from the car park (Access point) to reach the reservoir, passing under western hemlock trees. A 5km-long waymarked trail circuits the reservoir, or an easy route can be enjoyed among broadleaved trees near the Afon Crafnant. Another trail can be followed to Llyn Geirionydd in the next valley, passing by the remains of the Klondyke lead mill built in 1900 (SH7649962174 | ///evolution.seaside.cube). The views across Llyn Crafnant to the distant peaks of Crimpiau (475m) and Craig Wen (548m) are spectacular.

Looking across Llyn Crafnant [27] to the peaks of Crimpiau (475m).

HAFNA

LLANRWST, CONWY

MAP REF 28

Ownership: Public: Natural Resources Wales **Designations:** NP, SAC, SM, SSS **Area:** 36ha	**Forest type:** conifer **Forest location:** SH781601 **Explorer Map:** OL17 **Ease of access:** Easy/Moderate	**Access point:** SH7811460103 ///contexts.nurture.duplicity LL27 0JB

The old lead mine at Hafna is now encroached upon by spruce and pine trees, this being one of several sites that make up the huge Gwydir Forest [34]. The impressive, tiered remains of the nineteenth-century lead processing mill include bases of machinery, parts of the furnace, spoil heaps and an impressive chimney. Even from the car park the views across the forest are impressive, yet it is worth climbing the steps and exploring the old buildings and tunnelled passageways to gain an even better perspective of this fascinating site. Specialist rare plants and mosses, including alpine penny-cress and lead moss, now grow on the site's contaminated soils. A waymarked circular route (Moderate), the Hafna's Mining Trail, can be followed for 3.6km to explore more of the forest and other nearby mining sites.

An old passageway at Hafna [28].

Rowan foliage and flowers.

CORS BODGYNYDD

BETWS-Y-COED, CONWY   MAP REF 29

Ownership: Private: North Wales Wildlife Trust **Designations:** NP, SSSI **Area:** 1ha	**Forest type:** mixed **Forest location:** SH764595 **Explorer Map:** OL17 **Ease of access:** Easy/Moderate	**Access point:** SH7667959727 ///initiated.order.broached LL27 0JA

Large blocks of conifers surround this small nature reserve to the north of the main body of Gwydir Forest [34], centred on two lakes of Llyn Ty'n y Mynydd and Llyn Bodgynydd Bach, surrounded by wet acid bogland. Scattered broadleaves of birch, hawthorn and rowan grow in drier patches. The site is celebrated for hosting 16 species of dragonfly and damselfly, including black darters, black-tailed and keeled skimmers, and the scarce blue-tailed damselfly. A trio of distinctive birds favour the open treescape and woodland edges: cuckoo, and two remarkable crepuscular species, nightjar and woodcock. Although a wet site, a good path runs east–west through the reserve.

COED HEN DOETH

LLANBERIS, GWYNEDD   MAP REF 30

Ownership: Private **Designations:** **Area:** 7ha **Forest type:** temperate rainforest	**Forest location:** SH582593 **Explorer Map:** OL17 **Ease of access:** Moderate	**Access point:** SH5843259554 ///sectors.regulator.bottle LL55 4TY

Coed Hen Doeth is the surviving narrow central strip of upland sessile oak woodland among a larger forest dominated by Sitka spruce marked on maps as Coed Victoria. The current owner is an anaesthetist with the National Health Service, who proudly displays 'Guardian of Ancient Woodland' above her office door. The curious are informed that she is providing respite for pied flycatchers.

The wood teems with bryophytes and ferns. The distinctive large foliose lichen *Hypotrachyna endochlora* is found growing on nutrient-impoverished tree bark and mossy rocks. The site is rich with fungi in autumn. The entrance to the wood is almost opposite a hotel, which operates a private car park. The nearest available public parking is at the Snowdon Mountain Railway car park (charge), 200m to the west.

COED CAE HUDDYGL

BETWS-Y-COED, CONWY

MAP REF 31

Ownership: Public: Natural Resources Wales
Designations: NP, SM, SSSI
Area: 8ha

Forest type: mixed and temperate rainforest
Forest location: SH765583
Explorer Map: OL17

Ease of access: Difficult
Access point: SH7655358345
///airstrip.obstinate.equality
LL24 0DH

This steep-sided, densely wooded valley at the southern tip of Gwydir Forest in the Eryri/Snowdonia National Park is teeming with mosses and liverworts, which benefit from 170–190 days of rainfall each year. The sessile oak trees in this precious temperate rainforest follow the north bank of the Afon Llugwy, a tributary of the Afon Conwy, for 3.5km west of Betws-y-Coed, including the famous Swallow Falls. From the Access point, follow the Ty'n Llwyn trail down to the falls, whose mist sustains the bryophytes and ferns. The path can be narrow, with frequent sections with rocks and exposed tree roots, and is often steep. It passes through areas of natural and planted trees, including beech, oak and even a stand of 25 impressive hornbeam trees. The reward at the valley bottom is a dramatic view over the falls, with the advantage of being on the opposite bank to where throngs of tourists have paid to enjoy the same view. Climbers also come to tackle routes on the crag known as Craig Cae Huddygl, including 'Impossible Object' (grade VS 4c) and 'Gabriel's Horn' (HS 4b). A post-medieval lead mine can be discovered east of the falls (SH7719957698 | ///concerts.pampered.fights). The site can also be reached from Betws-y-Coed (see Gwydir Forest [34]).

Swallow Falls at Coed Cae Huddygl [31].

Below: Wood sorrel.

COED BRYN-ENGAN

CAPEL CURIG, CONWY   MAP REF 32

Ownership: Public: Natural Resources Wales **Designations:** IPA, NP, SSSI **Area:** 145ha	**Forest type:** temperate rainforest and mixed **Forest location:** SH716574 **Explorer Map:** OL17	**Ease of access:** Moderate/Difficult **Access point:** SH7165857709 ///hobbit.segmented.thunder LL24 0ET

About 20ha of this mixed woodland are considered temperate rainforest, rich with bryophytes, including shaded wood-moss, and the liverworts petty featherwort and greater whipwort, with many lichens also present including *Sticta limbata*.

There are two public footpaths through the wood. One heads south up the hill, the other stays near the valley bottom, heading east. Several informal routes can also be followed along forest rides. The Access point to the wood is behind (south of) the national outdoor training centre of Plas y Brenin. To reach the wood from the public car park (SH7203858242 | ///yours.toasted.insisting) while avoiding the main road, head north up the hill for 250m before following the path along the old leat towards the centre. From any high vantage point, enjoy dramatic views across to the Snowdon horseshoe.

Bryophytes growing on a larch branch.

ARTIST'S WOOD

BETWS-Y-COED, CONWY   MAP REF 33

Ownership: Public: Natural Resources Wales **Designations:** NP **Area:** 4ha	**Forest type:** conifer **Forest location:** SH776570 **Explorer Map:** OL17 **Ease of access:** Moderate	**Access point:** SH7779256864 ///king.irritate.otherwise LL24 0DA

Although only a minor area of conifers within the huge Gwydir Forest [34], Artist's Wood is fascinating due to long-term forestry research conducted on the site. Scientists are studying how Douglas fir responds in a forest structure with more than two distinct storeys. As a so-called 'plenter system', it has a wide range of differently sized and aged trees growing together. This makes management complex for the forester, and requires trees that are shade tolerant. Douglas fir does not thrive in deep shade, unlike some coniferous species like western hemlock, but with careful light management it can be encouraged to reproduce and grow very productively. Such forests are likely to be more resilient to environmental change than even-aged stands and able to support the future needs of society. Data from this site are being compared with another located in the Swiss Jura Mountains.

A coal tit calls from a beech tree.

GWYDIR FOREST

BETWS-Y-COED, CONWY   MAP REF 34

Ownership: Public: Natural Resources Wales	**Forest type:** mixed	**Access point:** SH7915056772
Designations: NP, SM, SAC, SSSI	**Forest location:** SH785584	///wedge.trap.enjoys
Area: 1,387ha	**Explorer Map:** OL17	LL24 0BB
	Ease of access: Easy/Moderate	

Gwydir Forest surrounds the town of Betws-y-Coed in the Eryri/Snowdonia National Park, and was one of the first 14 sites declared in 2020 as the Wales National Forest. While mostly coniferous, the forest is being actively managed and diversified with more broadleaves. Mining for lead and zinc were once important local industries, with building ruins, leats, reservoirs and spoil heaps waiting to be discovered throughout the forest, such as nearby Coed Mawr Pool lead mine (SH7811058434 | ///bookmark.rooftop.irrigate). Portions of the forest have special conservation designations, notably for the very tiny lead moss, which grows only on the most toxic areas of lead mining spoil heaps.

The Access point and the start to several waymarked routes is at Pont y Pair car park in town. Nearby, a section of boardwalk provides an easy start to exploring the fringes of the forest near the Afon Llugwy, including some very tall Douglas fir trees. Longer routes extend as far as Llyn y Parc, which makes a thrilling 10km circular route offering spectacular views of the forest and to the distinctive solitary peak of Moel Siabod (872m) to the south-west. Coed Cae Huddygl [31] lies 2km upstream and can also be accessed from the Access point by following the Eryri/Snowdonia Slate Trail.

Other sites detailed within Gwydir Forest include Artist's Wood [33], Cors Bodgynydd [29], Coed Cae Huddygl [31], Hafna [28] and Llyn Crafnant [27].

FFOS ANODDUN/FAIRY GLEN

BETWS-Y-COED, CONWY   MAP REF 35

| Ownership: Private
Designations: NP, SSSI
Area: 43ha
Forest type: broadleaved and temperate rainforest | Forest location: SH808536
Explorer Map: OL18
Ease of access: Moderate/Difficult | Access point: SH8108453534
///spoiled.screening.dangerously
LL24 0PN |

Ffos Anoddun/Fairy Glen is a narrow wooded gorge of the Afon Conwy. The attractive falls of Rhaeadr Y Graig Lwyd/Conwy Falls are worth a visit, but are much less dramatic during dry periods. The rocks and trees are clad in bryophytes and ferns, which is why some parts of the woods are considered to be temperate rainforest. The most accessible part of the mystical glen is on private land, accessed from the Conwy Falls Cafe. Parking (Access point) is free, but entry to the falls is charged at a turnstile. The circular one-way walk is quite short but is often steep, with exposed rocks and tree roots, and occasional steps and precipitous sides to the path.

ALWEN FOREST

CERRIGYDRUDION, CONWY   MAP REF 36

| Ownership: Public: Natural Resources Wales
Designations:
Area: 1,606ha | Forest type: conifer
Forest location: SH946545
Explorer Map: 264 and OL18
Ease of access: Moderate | Access point: SH9560752996
///casino.shelving.tonsils
LL21 9TT |

The vast conifer blocks of Alwen Forest are spit in two by a 5km-long reservoir, but thanks to a range of trails they are easy to explore. Some routes are more adventurous than others, especially a 15km-long trail that circumnavigates the reservoir, and the 23km-long Two-Lakes Trail, which links with Llyn Brenig to the east. Both routes pass through the heath and bogs of Hafod Elwy Moor NNR, where both red and black grouse can be seen, plus merlin and skylark. Alwin Forest contains a wide range of birdlife, including sparrowhawk and goldcrest, and all three native species of woodpecker.

TŶ MAWR WYBRNANT

PENMACHNO, CONWY   MAP REF 37

| Ownership: Private: National Trust
Designations: NP
Area: 5ha | Forest type: mixed
Forest location: SH770522
Explorer Map: OL18
Ease of access: Moderate | Access point: SH7713452369
///souk.yummy.drizzly
LL25 0HJ |

The bright waters of the Afon Wybrnant flow from a large (687ha) Sitka spruce plantation owned by Natural Resources Wales to reach this tranquil valley with mixed woods owned by the National Trust. In the heart of the valley is the sixteenth-century farmhouse and birthplace of William Morgan, who became a bishop and famously translated the Bible into Welsh. His life's work effectively helped preserve the Welsh language. Waymarked trails lead from the car park (Access point) to the farmhouse, while longer routes can be explored through the mixed woods.

Temperate rainforests

Wales is fortunate to have forests that are equally as special as tropical rainforests, only even rarer. These temperate rainforests, also known as Atlantic or Celtic rainforest in Wales, provide a unique habitat for wildlife, some of which is internationally significant. The high annual rainfall, relatively mild winters and low pollution levels found in the western regions of Wales provide the essential ingredients of a temperate rainforest. They are often characterised by prolific growth of bryophytes (see pp.116–17) like tree lungwort, while providing habitat for iconic birds that are uncommon across most of Britain, notably pied flycatcher, redstart, wood warbler and lesser spotted woodpecker.

The greatest number of temperate rainforest sites are found in the North-West region, mostly within the Meirionnydd Oakwoods and Bat Sites SAC, including: Coed Aberartro [60], Coed Llechwedd [54], Coed Cymerau Isaf [44], Coed Felenrhyd and Llennyrch [48], Coed Lletywalter [57] and Coed Garth Gell [65]. Sites in other regions include Gwenffrwd-Dinas [201] (South-West), Carngafallt [167] (Central-East) and Coed Cwm Einion [104] (Central-West).

All these precious sites are mere fragments of natural woodland that once colonised much of western Wales after the retreat of the ice sheets at the end of the last glacial period. Those we see today exist mainly because they have remained inaccessible to grazing livestock and deer, or are difficult to cut regularly for use as timber or fuel in industry. In particular, the sessile oak trees, which are the dominant tree species, were managed for their bark and used in the tanning industry (see p.14). Many areas were felled to make way for plantation forests in the twentieth century, now being addressed by strategies to restore Plantations on Ancient Woodlands (PAWS) sites (see pp.188–89).

Recently, a new collaborative project has been launched, which is seeking to regenerate temperate rainforests across a number of sites in Wales. Coedwigoedd Glaw Celtaidd Cymru (Celtic Rainforests Wales) is a consortium of public and private partners working at scale to improve conditions of existing fragments of rainforest with the long-term ambition of expanding rainforest areas. One of its main tasks is to limit unwanted grazing, both by controlling population numbers of wild deer, and by using hardy ponies and long-horned cattle to graze some sites, as they eat plants more selectively than sheep. Another major challenge for landowners is the control of exotic and invasive species, especially common rhododendron, *Rhododendron ponticum*. This type of rhododendron smothers the soil with deep shade and prevents native ground flora from growing, affecting the entire habitat if allowed to grow unchecked.

Exploring temperate rainforest sites can be a challenge, especially given the steep and often remote terrain where they can be found growing today. Leaving the safety of a footpath can be hazardous, given the luxuriant plant growth and preponderance of large boulders, yet it is these natural barriers that contribute to their biological richness and they are best left undisturbed. Instead, carry binoculars and take time to stop, listen and admire their beauty.

Opposite: tree lungwort.

NANTGWYNANT

BETHANIA, GWYNEDD | MAP REF 38

Ownership: Private: National Trust, and Public: Natural Resources Wales **Designations:** IPA, NNR, NP, SAC, SSSI	**Area:** 63ha **Forest type:** temperate rainforest **Forest location:** SH626508 **Explorer Map:** OL17 **Ease of access:** Moderate/Difficult	**Access point:** SH6278250682 ///discloses.reboot.massaged LL55 4NR

Situated in the heart of the Eryri/Snowdonia National Park, the Access point is used by walkers starting the Watkin Path, which is among the most rewarding routes to ascend Snowdon (1,085m). While most visitors will have their eyes on the highest peak in Wales, a group of three sessile oak woods in the valley are fragments of temperate rainforest. North of the valley bottom and the A498, Coed-yr-allt and Hafod-y-Llan fall partly within the IPA, NNR, SAC and SSSI, while Coed Eryr lies otherwise undesignated to the south. Bryophytes and mosses clad the trees, including the rare flagellate feathermoss and greater streak-moss. Public footpaths provide access to all three sites, but the easiest route is through the National Trust sites at Hafod-y-Llan and along the relatively level path north-west of Llyn Gwynant.

BEDDGELERT FOREST

BEDDGELERT, GWYNEDD | MAP REF 39

Ownership: Public: Natural Resources Wales **Designations:** NP **Area:** 442ha	**Forest type:** conifer **Forest location:** SH566500 **Explorer Map:** OL17 **Ease of access:** Moderate	**Access point:** SH5740050300 ///announced.boom.sonic LL54 6TW

With occasional dramatic views to the peak of Snowdon, panoramas across the vast conifer blocks, and a secretive lake, it is no wonder that this large forest is popular with walkers and cyclists. A single waymarked circular trail can be followed on forest roads for 4km, while mountain bikers are offered an alternative route. Llyn Llywelyn provides welcome diversity, while the mews of buzzard and whistle of narrow-gauge engines on the Welsh Highland Railway create an enigmatic soundscape.

COED BWLCH-DERW

BEDDGELERT, GWYNEDD | MAP REF 40

Ownership: Private: National Trust **Designations:** IPA, NP, SAC, SSSI **Area:** 5ha	**Forest type:** temperate rainforest **Forest location:** SH618493 **Explorer Map:** OL17 **Ease of access:** Moderate	**Access point:** SH6113749352 ///forge.handbag.ramp LL55 4NG

Enjoy a peaceful easy stroll along the southern shore of Llyn Dinas on a well-surfaced trail, or climb through the trees for a closer look at this fragment of temperate rainforest. The going can be strenuous, but looking carefully at the contorted branches of the sessile oak and rowan trees might reveal the feather-like fronds of the many-leaved pocket moss and several unusual lichens. The views across the lake to the peak of Snowdon are magnificent. From the lay-by (Access point) walk a short distance to the east and cross the road to reach a kissing gate and the Cambrian Way.

COED CRAFLWYN AND DINAS EMRYS

BEDDGELERT, GWYNEDD — MAP REF 41

Ownership: Private: National Trust **Designations:** HPG, NP, SAC, SM, SSSI **Area:** 27ha	**Forest type:** mixed **Forest location:** SH597490 **Explorer Map:** OL17 **Ease of access:** Moderate	**Access point:** SH5995948972 ///relished.sugars.notes LL55 4NF

Two distinct areas of woodland can be explored on the extensive Craflwyn and Beddgelert estate in the glorious valley of Nant Gwynant. Coed Craflwyn (not to be confused with site 13 near Rowen which has the same name) is a mixed woodland of Scots pine and broadleaves west of the hall and car park (Access point). To the east on a prominent hill is a broadleaved woodland dominated by sessile oak, areas of which are considered temperate rainforest.

The hill, named Dinas Emrys, became a defensive settlement in the Roman era, if not earlier, followed by successive generations of people taking advantage of the clear views across the valley. The early walls are indistinct, but the square footprint of a medieval stone tower can be found.

In the legend of King Arthur, the hill is the setting of a famous argument between fifth-century warlord Vortigern and the youthful wizard, Merlin. The warlord had retreated to the hill fort to escape the Romans, but every day the buildings his army constructed would collapse. Finally, Vortigern was told to seek the advice of a young man not conceived by mortals. After a long search, his men finally found the young Merlin, who explained that the buildings were unstable because two vermes (dragons) were battling in a cavern below. Although the White Dragon of the Saxons was winning the battle, it would nonetheless soon be defeated by the Welsh Red Dragon. After Vortigern's death, the war leader Emrys Wledig (Aurelius Ambrosius in English) took over the site, hence its name.

A range of waymarked trails are provided. Those in the valley bottom are easy-going, but climbing Dinas Emrys is quite strenuous.

COED DOLFRIOG

NANTMOR, GWYNEDD — MAP REF 42

Ownership: Private **Designations:** IPA, NP, SAC, SSSI **Area:** 43ha	**Forest type:** mixed and temperate rainforest **Forest location:** SH610459 **Explorer Map:** OL17	**Ease of access:** Moderate **Access point:** SH6202246749 ///soils.flattens.expressed LL48 6SN

Enjoy tranquillity and solitude at Dolfriog, even though you will be close to busy tourist areas in the Eryri/Snowdonia National Park. From a small lay-by (Access point) on the single-track road, walk south for 150m and follow a track down the side of a small plantation towards a ford over the beautiful Afon Nanmor. After a short, steep climb, follow the footpath, which runs southwards keeping a more or less level route along the edge of the wood. On the sessile oak, rowan and downy birch, admire the bryophytes that mark this as a temperate rainforest, including Killarney featherwort and greater streak-moss.

COED ELERNION

TREFOR, GWYNEDD
MAP REF 43

Ownership: Private: Woodland Trust
Designations: NL, SSSI
Area: 21ha

Forest type: broadleaved
Forest location: SH378460
Explorer Map: 254
Ease of access: Easy/Moderate

Access point: SH3815746130
///fraction.animate.flashback
LL54 5HH

Even though it grows under the northern flanks of the dominant peaks of the Yr Eifl hills, this small broadleaved site is prominent because the Llŷn Peninsula is otherwise quite depauperate of woodland. It is highly diverse, with ash, downy birch and sessile oak dominating in dry areas, while alder and grey willow favour wetter areas. Three flower-rich meadows are hidden within the wood, and there are areas of scrub and marshland. A circular path can be followed through the site, which is fairly level, but the paths are often quite wet and can become muddy.

Parking is limited near the entrance, which comprises a field gate and squeeze posts.

COED CYMERAU-ISAF

RHYD-Y-SARN, GWYNEDD
MAP REF 44

Ownership: Public: Natural Resources Wales
Designations: IPA, NNR, NP, SAC, SSSI

Area: 26ha
Forest type: temperate rainforest
Forest location: SH690430
Explorer Map: OL18

Ease of access: Moderate
Access point: SH6939443122
///apes.bulbs.quite
LL41 4NT

The heart of this NNR is in the deep valley sides of the Afon Goedol, and mostly quite inaccessible. Like its neighbouring sites of Coed Llyn Mair [46] and Coed Felenrhyd and Llennyrch [48] within the Meirionnydd Oakwoods IPA, the sessile oak trees and humid atmosphere provide the perfect habitat for hosts of ferns and bryophytes.

There are three access options (all with parking close by), which make it possible to enjoy the attractive waterfalls at the northern tip of the reserve, which otherwise remain beyond reach. Nonetheless, the surrounding woodland is beautiful and harbours redstart, pied flycatcher and nuthatch. The closest and easiest route from the Access point still involves some climbing and sections of steps. A trail can be followed through trees to the open ground of Parc Cymerau-isaf, with views across the valley reserve, while a branch of the path descends to the falls.

It is also possible to approach from the south (SH6907342305 | ///balancing.point.finer), which involves a climb from the start through attractive woodland. Alternatively, from the north (SH6933943996 | ///blows.wins.series), follow a footpath south through a conifer plantation for 1km to reach the falls.

An unfurling fern frond.

COED HAFOD Y LLYN

TAN-Y-BWLCH, GWYNEDD

MAP REF 45

Ownership: Private: Woodland Trust	**Forest type:** mixed	**Access point:** SH6521441356
Designations: IPA, NP, SAC, SSSI	**Forest location:** SH650411	///finally.jugs.refilled
Area: 31ha	**Explorer Map:** OL18	LL41 3AQ
	Ease of access: Easy/Moderate	

A path above the Access point climbs into the temperate rainforest of Coed Llyn Mair [46], but cross the road to enter Coed Hafod y Llyn and enjoy a fairly level 4km-long circular trail skirting the southern shores of the llyn. Once part of the Plas Tan y Bwlch estate, the lake was created in 1889 to provide hydroelectric power, and its native trees cleared to grow softwood timber. Pockets of conifers remain, especially near the lake, including some large specimens of grand fir. Oak trees clad in ferns, mosses, lichens and liverworts indicate temperate rainforest conditions. The Ffestiniog Railway winds its way higher up the slopes, the whistle of its narrow-gauge engines echoing across the valley.

The Ffestiniog Railway line passes through Coed Hafod y Llyn [45].

It is possible to alight from the stations at either Tan-y-Bwlch or Plas Halt to reach the wood, or even walk through the trees between the two.

A view across the lake at Coed Hafod y Llyn [45].

COED LLYN MAIR

TAN-Y-BWLCH, GWYNEDD

MAP REF 46

Ownership: Public: Natural Resources Wales **Designations:** IPA, NNR, NP, SAC, SSSI	**Area:** 5ha **Forest type:** temperate rainforest **Forest location:** SH651414 **Explorer Map:** OL18	**Ease of access:** Moderate **Access point:** SH6521441356 ///finally.jugs.refilled LL41 3AQ

An entrance gate to Coed Llyn Mair [46] under a spreading sessile oak.

An 8km-long stretch of the north side of the Afon Dwyryd in the Vale of Ffestiniog harbours an extraordinary cluster of sessile oak woods, designated as a large (354ha) SSSI named Coedydd Dyffryn Ffestiniog (North). A substantial proportion is also designated as the Coedydd Maentwrog NNR, containing at least four separate named woods: Coed Llyn Mair to the west of the B4410, and a group to the east (Coed Ty Coch, Coed Bronturnor and Coed Glanrafon). The Ffestiniog Railway runs along its northern fringe, providing an unforgettable experience and a comfortable means to view the beautiful woods below. Passengers can alight at either of two holts at opposite ends of the wooded valley (Tan-y-Bwlch to the west, and Dduallt to the east) to explore the wooded valley on foot. However, as walkers following the Cambrian Way discover while following the trail that also runs through much of the valley, the going is often demanding.

Coed Llyn Mair provides the most accessible option among these spectacular woods. A 600m circular trail can be followed from the Access point, which is steep in places and involves several flights of steps. It is also possible to walk through the trees between Tan-y-Bwlch and Plas Halt stations. In summer, listen for wood warbler in between the whistle of the steam train. Lesser horseshoe bats emerge at dusk to feed on the rich invertebrate life. This temperate rainforest supports a huge diversity of mosses, ferns, liverworts and lichens. In damp areas, especially on willow and hazel stems, look for the brown lichen species *Sticta sylvatica*. The creeping liverwort, straggling pouchwort, can be found on rocks, its conspicuous flat green leaves often flecked with gold and brown.

Coed Hafod y Llyn [45] can be accessed from the same Access point.

CORS-Y-SARNAU

SARNAU, GWYNEDD — MAP REF 47

Ownership: Private: North Wales Wildlife Trust **Designations:** SSSI **Area:** 15ha	**Forest type:** broadleaved **Forest location:** SH973393 **Explorer Map:** OL18 **Ease of access:** Moderate	**Access point:** SH9733339394 ///manliness.lost.positions LL23 7LF

The full dynamic nature of succession can be seen in slow-motion action on this fascinating wetland nature reserve. Technically known as a hydrosere, the shallow lake and surrounding boggy ground (*cors* in Welsh) is slowly converting from open water to 'climax' woodland. As silt gradually builds up, marshland begins to develop and eventually, perhaps millennia later, woodland will take over.

The site is rich with all forms of wildlife, from grass snake and common lizard, to several species of warbler, snipe and woodcock. Insectivorous sundews thrive on the damp peat, while alder and willows form the carr (wet woodland).

The reserve entrance (Access point) is on the south side of the busy A494. Park in the village and approach warily, as there is no pavement along the road and crossing can be difficult. The site is very level but wet underfoot.

COED FELENRHYD AND LLENNYRCH

MAENTWROG, GWYNEDD — MAP REF 48

Ownership: Private: Woodland Trust **Designations:** IPA, NNR, NP, SAC, SSSI	**Area:** 310ha **Forest type:** temperate rainforest **Forest location:** SH654390 **Explorer Map:** OL18	**Ease of access:** Moderate/Difficult **Access point:** SH6532139547 ///blank.inflation.shall LL41 4HY

On the south side of the Vale of Ffestiniog, opposite Coed Llyn Mair [46], lies another large area of temperate rainforest. Its sessile oak trees, with occasional rowan and hazel, plus alder in wet areas, stretch along the stunning gorge of the Afon Prysor (which flows from Llyn Trawsfynydd). It forms part of the Meirionnydd Oakwoods IPA.

The woods are particularly valuable for bryophytes, which prosper in the magical mist-laden atmosphere. Among them the blue-grey foliose lichen *Hypotrachyna laevigata*, and the barnacle lichen, whose crusty grey fruiting bodies grow on tree branches resembling the sea crustacean of the same name. Some 200 liverwort species have been recorded locally, including pale scalewort, long-leaved, toothed, and pointed pounceworts, and western earwort. It is also the only known location in Britain for the rare prostrate feathermoss, among a number of other rare mosses, such as shaded wood, silky swan-neck and club pincushion. Ferns also thrive here, notably narrow buckler, oak and royal. Flowering plants include wood sorrel and globeflower. Look for dipper near the water, and nuthatch and treecreeper scaling the mossy tree trunks.

The entrance to the woods (Access point) is immediately west of the bridge over the Afon Prysor, near the entrance to the Maentwrog power station. Park in a lay-by a short distance to the north-east (SH6543539679 | ///haircuts.fallen.mats).

GARN BODUAN

NEFYN, GWYNEDD

MAP REF 49

Ownership: Private **Designations:** NL, SM **Area:** 81ha **Forest type:** conifer	**Forest location:** SH313395 **Explorer Map:** 253 **Ease of access:** Moderate/Difficult	**Access point:** SH3086840413 ///spare.rural.entire LL53 6HH

Together with the three peaks of Yr Eifel, Garn Boduan is one of the prominent hills on the otherwise gently rolling landscape of the Llŷn Peninsula. A previous crop of conifers on its flanks was felled in the 2010s and replanted with Sitka spruce and larch, and with lodgepole pine near the nutrient-poor and exposed summit. Some broadleaves have also been planted, especially oak, birch and holly. Its rocky summit (279m) rises above the trees, providing breathtaking views, taking in the sandy beach at Nefy, and the whole Llŷn Peninsula. To the north, Mynydd Twr/Holyhead Mountain is visible on Holy Island, while to the east the distant high peaks of the Snowdon range, and to the south extensive views along the coast to Bae Abermaw/Barmouth Bay and the distinctive summit of Cadair Idris. No wonder the large (10ha) prehistoric hill fort on its summit once sheltered 170 roundhouses. Walk from the town, or use a free car park (Access point). Walk up Y Fron lane to reach the forest (SH3133539979 | ///finger.station.warthog). The walk is moderately strenuous and requires some navigation skills, and towards the summit involves a steep scramble.

Looking north-east from the summit of Garn Boduan [49].

GWAITH POWDWR

PENRHYNDEUDRAETH, GWYNEDD MAP REF 50

| Ownership: Private: North Wales Wildlife Trust Designations: IPA, SAC, SSSI Area: 13ha | Forest type: broadleaved Forest location: SH618389 Explorer Map: OL18 Ease of access: Easy | Access point: SH6163138858 ///lavender.putty.nanny LL48 6LT |

Nesting pied flycatcher, redstart and tree pipit, and hordes of butterflies and flowering plants belie an incendiary past for this fascinating nature reserve. For 130 years the steep-sided valley provided an ideal location for an explosives factory, producing munitions for the British Army, including 17 million grenades during the Second World War. Fragments of the site fall under the Meirionnydd Oakwoods IPA. The areas where woodland meets surrounding heathland are perfect locations to spot nightjar, while colonies of lesser horseshoe bat nest in old buildings and tunnels.

Many parts of the reserve are easy to explore, thanks to the old metalled roads. Enjoy wonderful views of the Dwyryd estuary from higher open ground. There is room for a couple of cars near the reserve entrance on the Cooke's industrial estate (Access point).

An oak tree grows from crags at Gwaith Powdwr [50].

TRAETH GLASLYN

MINFFORDD, GWYNEDD MAP REF 51

| Ownership: Private: North Wales Wildlife Trust Designations: SSSI Area: 2ha | Forest type: broadleaved Forest location: SH586381 Explorer Map: OL18 Ease of access: Easy/Moderate/Difficult | Access point: SH5846237941 ///added.happier.brew LL49 9AW |

This thin strip of wet willow woodland lies alongside the mudflats and salt marsh of the Afon Glaslyn estuary, managed as a nature reserve. Parking is available immediately north-east of the entrance to the reserve (Access point). Look for the pedestrian gate on the cycle path, just beyond a wooden arch. It is a short walk to a bird hide with wonderful views across the estuary and its wading birds. Paths can be followed further into the reserve, but take caution because they regularly flood due to rising tides, and should not be deviated from, especially at the reserve's northern end. Look for willow warbler in the trees, and take binoculars to spot black-tailed godwit, curlew and redshank feeding at low tide.

THE GWYLLT

PORTMEIRION, GWYNEDD MAP REF 52

Ownership: Private: Portmeirion Estate
Designations:
Area: 38ha
Forest type: mixed
Forest location: SH586370
Explorer Map: OL18
Ease of access: Moderate
Access point: SH5907537414
///drape.artichoke.hissing
LL48 6ER

Surrounding the popular tourist destination of Portmeirion village lies a large subtropical woodland bursting with exotic tree species. A very extensive network of trails can be followed to explore the trees, glades and dramatic coastline. Many of the trees were planted before Clough Williams-Ellis started work collecting and reconstructing endangered buildings to create the unique eco-friendly village. Victorian owners (including telegraph pioneer William Fothergill Cooke) planted giant redwood, monkey puzzle and exotic pines. Later, large collections of rhododendron were planted, along with magnolia and ginkgo. The 'dancing tree' near the Japanese lake is an unusually large New Zealand privet.

The peninsula is privately owned, and access is charged, both for parking (Access point) and to walk through the woodland. Paths are often steep and can be slippery when wet.

The Gwyllt [52] and colourful buildings of Portmeirion.

ABERHIRNANT

BALA, GWYNEDD MAP REF 53

Ownership: Public: Natural Resources Wales
Designations: NP
Area: 2,059ha
Forest type: conifer
Forest location: SH957326
Explorer Map: OL18 and OL23
Ease of access: Moderate
Access point: SH9573032689
///amplified.centuries.evaded
LL23 7EY

This huge conifer plantation forest fans out from Cwm Abernant, covering the catchments of the Afon Glyn, which flows west into Llyn Tegid/Bala Lake and the Hirnant, which heads east to the Afon Dyfrdwy/River Dee. Forest rides offer plenty of choice for explorers, but given the long distances involved are perhaps best enjoyed on an all-terrain bicycle. The Access point is a small car park with a picnic site, reached along a minor road linking Llyn Tegid /Bala Lake with Llyn Efyrnwy/Lake Vyrnwy [138]. South of here lining the sides of the road are some very large Sitka spruce trees (SH9472526444 | ///complains.blues.offstage), some that peak at more than 40m tall.

COED LLECHWEDD

HARLECH, GWYNEDD — MAP REF 54

Ownership: Private: Woodland Trust **Designations:** IPA, NP, SAC, SSSI **Area:** 26ha	**Forest type:** broadleaved and temperate rainforest **Forest location:** SH592317 **Explorer Map:** OL18	**Ease of access:** Moderate **Access point:** SH5902831717 ///swordfish.trainer.glue LL46 2UU

Occupying a steep north-west-facing slope near the impressive castle at Harlech, this beautiful and diverse ancient woodland is unspoilt and rich with wildlife. Considered one of the best temperate rainforest sites in Wales, it falls under the Meirionnydd Oakwoods and Bat Sites SAC. Ash, sycamore and wild cherry give way to sessile oak, birch and rowan on higher slopes. Look for soft shield-fern and broad buckler fern, and bryophytes on the tree's branches, where you may also be lucky to see a lesser spotted woodpecker. There is also a small heronry within the wood. At its upper reaches the wood gives way to scrub and scattered trees, known as 'ffridd', where extensive views can be enjoyed of the town, castle and Bae Tremadog/Tremadoc Bay.

A bridleway leads into the wood from the B4573 (Access point) but there is no parking here. Leave a car in nearby Harlech and follow a footpath from the town centre to avoid walking along most of the roads.

COED DOLFUDR

DOLHENDRE, GWYNEDD — MAP REF 55

Ownership: Private **Designations:** NP, SAC, SPA, SSSI **Area:** 21ha **Forest type:** temperate rainforest	**Forest location:** SH830320 **Explorer Map:** OL23 **Ease of access:** Difficult	**Access point:** SH8531930844 ///scrum.grumbles.rewrites LL23 7SY

This is a seldom-visited wood at the east of the Eryri/Snowdonia National Park, lying within a multi-designated area known as Migneint-Arenig-Dduallt, whose high lands include the second largest blanket bog in Wales. Some of its lower slopes include fragments of upland sessile oak woods, such as this site on the steep-sided western side of the Afon Lliw. A bridleway passes through the lower fringes of the wood, but exploring the ground above is a challenge given the steep and rocky terrain. Trees and boulders are clad in mosses and lichens, and in between harbour ferns, especially hard fern. From the Access point (a small car park next to the road bridge), walk south-west along the road to a junction, then head west for 500m along a quiet lane, climbing all the way. On the steep, open slopes above lies Castell Carn Dochan, a castle dating from the early thirteenth century. Its ruins are challenging to explore, but include impressive remains of a main wall with three towers, while the views are superb, making this a worthwhile diversion. After the metalled road turns into a rough track, it passes through the conifers of Coed Bryn Bras, with Coed Dolfudr found 1km beyond.

Polypody ferns.

COED CRAFNANT

PENTRE GWYNFRYN, GWYNEDD    MAP REF 56

Ownership: Private: North Wales Wildlife Trust **Designations:** ISA, NP, SAC, SSSI **Area:** 49ha	**Forest type:** temperate rainforest **Forest location:** SH619288 **Explorer Map:** OL18 **Ease of access:** Difficult	**Access point:** SH6165328954 ///running.poets.horn LL45 2PH

Halfway between Wales' best-kept hillwalking secret in the Rhinogs, and the coastal town of Harlech, the remote valley of the Afon Artro harbours a significant tract of temperate rainforest. Together with neighbouring woods Coed Dolbebin and Coed Dolwreiddiog, the reserve at Coed Crafnant forms 127ha of stunning sessile oak woodland, rich with bearded lichens, liverworts (including deceptive featherwort), mosses and ferns. Among the ancient trees, multitudes of nest boxes have been erected to boost populations of pied flycatcher, while nuthatch, redstart, tawny owl and wood warbler are among a large number of bird species to be seen or heard. Bluebells carpet the site in spring, and the yellow flowers of the semi-parasitic common cow-wheat brighten the fringes of the wood in summer.

The nearest entrance to the reserve is the Access point given, where it is just possible to park a single car by the side of the road. Alternatively, a small car park lies 600m to the north (SH6196129538 | ///vague.blacken.faded). The site is steep and paths rough, and there is unlikely to be a mobile phone signal.

An oak tree welcomes visitors to Coed Crafnant [56].

Looking south-east through the oak trees of Coed Lletywalter [57] towards the peak of Moelfre (589m).

COED LLETYWALTER

PENTRE GWYNFRYN, GWYNEDD   MAP REF 57

Ownership: Private: Woodland Trust
Designations: IPA, NP, SAC, SSSI
Area: 38ha

Forest type: temperate rainforest
Forest location: SH599276
Explorer Map: OL18
Ease of access: Moderate

Access point: SH6024727496
///digits.motivator.apes
LL45 2PB

Along with nearby Coed Aberartro [60] and Coed Hafod y Bryn [61], Coed Lletywalter is part of the Meirionnydd Oakwoods and Bat Sites SAC, and a group of sites recognised for their extraordinarily high conservation value. The wood is mostly sessile oak with occasional ash, beech, birch, sycamore and willow, with a healthy regenerating understorey of young trees, especially holly. None of the trees are very old, suggesting that timber from the woodland was probably harvested during the Second World War. Tree lungwort is found occasionally, but many trees and rocks are clad in other temperate rainforest lichens such as smooth loop lichen, plus mosses and liverworts. In autumn, look for penny bun fungus. Treecreeper and nuthatch are resident all year round, joined during spring and summer by pied flycatcher, redstart and wood warbler. This is an attractive and varied site to explore, with rocky knolls and boulder-strewn slopes, a small lake and stream, and an abandoned farmstead. A circular route through part of the site is waymarked, which includes some steep sections with steps. There is room for a single car to park in a small lay-by next to the wood entrance (Access point).

COED Y BRENIN FOREST PARK

DOLGELLAU, GWYNEDD

MAP REF 59

Ownership: Public: Natural Resources Wales **Designations:** NP **Area:** 2,065ha	**Forest type:** conifer **Forest location:** SH723268 **Explorer Map:** OL18 and OL23 **Ease of access:** Easy/Moderate	**Access point:** SH7233126888 ///merge.spaceship.plantings LL40 2HZ

Coed y Brenin is one of the first sites designated as a National Forest site in Wales. It was also the first purpose-built mountain biking centre in Britain, and continues to host world-class mountain bike trails and competitions. There are plenty of options for walkers too. Three waymarked trails start from the main car park and visitor centre (Access point), where there is parking charge.

An accessible trail reaches the Afon Eden, another extends to the Cefndeuddwr viewpoint and picnic area. The most strenuous option provides an attractive route to the confluence of two rivers and the impressive waterfalls of Pistill Cain and Rhaeadr Mawddach. There are also running trails and two geocaching routes.

While its large conifer blocks are perfect hunting territory for goshawk, the remoter areas also provide lekking sites for the rare black grouse. Large mounded nests of hairy wood ant may also be seen.

There are several alternative car parks in the forest, which, having fewer facilities, are free to use. From the Pandy car park (SH7447022447 | ///aged.restriction.scored), follow an easy trail through a collection of exotic trees at the Coed y Brenin Forest Garden.

The adventurous can even stay overnight in one of Wales' nine bothies at Penrhos Isaf (SH7377623873 | ///zone.lofts.bombshell).

Main image: Looking across the rolling forested hills of Coed y Brenin [59] from the Precipice Walk at Nannau [64].

FFYNNON SAINT

RHIW, GWYNEDD   MAP REF 58

Ownership: Private: National Trust	Forest location: SH242293	Access point: SH2415728561
Designations: NL	Explorer Map: 253	///overgrown.increased.feuds
Area: 42ha	Ease of access: Moderate	LL53 8AA
Forest type: conifer		

Christians once flocked to Bardsey Island during the medieval period, and on their pilgrimage through the Llŷn Peninsula would stop at holy wells to cleanse themselves. Ffynnon Saint is one of the best-preserved wells, measuring almost 3m square and featuring thick walls with two sets of steps (SH2420029470 | ///reclaimed.acting.smug). From the car park (Access point), cross the road and follow the track north. It passes through blocks of Sitka spruce, which are unremarkable, except for the shelter they now provide to visitors prepared to climb the short slope to reach the well, which is a very tranquil place.

COED ABERARTRO

PENTRE GWYNFRYN, GWYNEDD   MAP REF 60

Ownership: Private: Woodland Trust	Forest type: temperate rainforest and broadleaved	Ease of access: Easy/Moderate
Designations: IPA, NP, SAC, SSSI	Forest location: SH599267	Access point: SH6010926823
Area: 27ha	Explorer Map: OL18	///libraries.saloons.shallower
		LL45 2NA

This wood once formed part of the grounds of Aber Artro Hall, the birthplace of poet Patrick Shaw-Stewart (1888–1917) who wrote 'Achilles in the Trench'. The remains of an arboretum and other exotic planted trees are present, together with sycamore and beech. The majority of the site is semi-natural ancient woodland, mostly comprising sessile oak with ash and hazel. The entire wood is designated an SAC, while the area east of the minor road is an SSSI. Liverworts including prickly and spotted featherworts have been discovered near the river, and the more common western pouncewort on higher areas. In spring, the display of bluebells is particularly spectacular, and can even be enjoyed from the roadside. The site can also be reached via a pedestrian bridge, which crosses the Afon Artro at Pentre Gwynfryn, or as part of an enjoyable longer walk from Llanbedr. Coed Lletywalter [57] lies on the opposite side of the valley.

Young hazelnuts developing on a hazel sprig.

COED HAFOD Y BRYN

LLANBEDR, GWYNEDD MAP REF 61

Ownership: Public: Snowdonia National Park Authority **Designations:** IPA, NP, SAC **Area:** 7ha	**Forest type:** broadleaved **Forest location:** SH586264 **Explorer Map:** OL18 **Ease of access:** Easy/Moderate	**Access point:** SH5865926533 ///defenders.member.wrong LL45 2ND

Overlooking the town of Llanbedr, this small wood is managed by a local community group, the Friends of Llanbedr Woodlands. Under the sessile oak, ash, sweet chestnut and beech trees, carpets of bluebells light up the woodland floor in spring. Despite its elevated position, the nearby sea at Bae Tremadog/Tremadoc Bay can be more sensed than seen, with the trees hampering any view from within the wood. Along with nearby Coed Aberartro [60] and Coed Lletywalter [57] the wood forms part of the Meirionnydd Oakwoods and Bat Sites SAC. The local council have developed a trail that links all three woodlands in the Afon Artro valley, extending all the way to another, Coed Crafnant [56], 5km further upstream.

The entrance to Coed Hafod y Bryn [61].

COED GANLLWYD

GANLLWYD, GWYNEDD

MAP REF 62

Ownership: Private: National Trust Designations: IPA, NNR, NP, SAC, SSSI Area: 24ha	Forest type: temperate rainforest Forest location: SH723244 Explorer Map: OL18 Ease of access: Moderate	Access point: SH7267624352 ///everybody.caged.pipe LL40 2TF

Coed Ganllwyd is an ancient woodland oasis at the southern end of the enormous sprawling conifers of Coed y Brenin Forest Park. The NNR area surrounds the eastern and lower reaches of the Afon Gamlan before its confluence with the Afon Mawddach. Like other woods that fall under the Meirionnydd Oakwoods IPA, its trees host rare bryophytes, while its SAC status is awarded due to its provision of perfect habitat and maternity roosting sites for the lesser horseshoe bat.

From the car park on the A470 (Access point), climb the waymarked trail into the woods, which is often steep and rocky. Keep a look out for pied flycatcher in the sessile oak trees and black alder, and dipper feeding in the water. About 1km to the west are the spectacular Rhaeadr Ddu/Black Falls, consisting of two waterfalls cascading more than 18m.

Grey wagtail perched in a black alder.

COED CORS-Y-GEDOL

TAL-Y-BONT, GWYNEDD

MAP REF 63

Ownership: Private Designations: IPA, NP, SAC, SSSI Area: 57ha Forest type: temperate rainforest	Forest location: SH598224 Explorer Map: OL18 Ease of access: Easy/Moderate	Access point: SH5897321807 ///elbow.local.undercuts LL43 2AN

Sessile oak and alder trees shade the banks of the tumbling Afon Ysgethin, from 200m above sea level (surrounded by the extensive remains of a Bronze Age settlement) to the outskirts of Tal-y-bont near the coast. Every branch and stem is clad in bryophytes, notably several species of leafy liverworts including tree lungwort, straggling pouchwort and western earwort. Follow the path that begins at the back of a small car park (Access point) to enjoy this very scenic woodland.

NANNAU

LLANFACHRETH, GWYNEDD   MAP REF 64

Ownership: Private: Nannau Estate Designations: HPG, NP, SSSI Area: 18ha	Forest type: temperate rainforest Forest location: SH736208 Explorer Map: OL18 Ease of access: Moderate	Access point: SH7456721181 ///tensions.smirking.caveman LL40 2NG

The twelfth-century estate of Nannau has a rich cultural history. According to legend, in 1402 Owain Glyndŵr's cousin Hywel Sele, who then owned the land, attempted to shoot the Welsh military leader during a hunting trip, but the hero slayed his treacherous relative before hiding his body in a hollow oak tree.

Beautiful sessile oak woods on the Nannau estate surrounding Llyn Cynwch – a reservoir providing Dolgellau with drinking water – are considered fine examples of temperate rainforest. Notable bryophytes include *Sticta sylvatica*, deceptive and spotty featherworts, and tree lungwort. Keep a look out for treecreeper among the mossy oak trees.

Public access has been granted by the owners to the wooded area on the western shore of the llyn since 1890, where the path that passes through it forms part of the famous Precipice Walk. The 5km circuit provides superlative views of the mountains of North and Mid Wales, from Eryri (Snowdon) to Cadair Idris, and across the extensive plantations of Coed y Brenin Forest Park [59]. The route is well named, given the steep hillside, but the path contours around the hill, requiring little climbing. Nonetheless, given its exposure, visitors should be equipped for hillwalking. Scattered Scots pine, birch and rowan trees cling to the steep north-west slopes. The Access point is the Saith Groesffordd car park.

Treecreeper.

COED GARTH GELL

BONTDDU, GWYNEDD MAP REF 65

Ownership: Private: RSPB
Designations: IPA, NP, SAC
Area: 37ha
Forest type: temperate rainforest
Forest location: SH684199
Explorer Map: OL18 and OL23
Ease of access: Moderate
Access point: SH6780118974
///detergent.cape.kingdom
LL40 2UL

Growing on the west side of the Afon Cwm-mynach, a tributary of Afon Mawddach, this fabulous temperate rainforest hosts many veteran oak trees. Their branches and stems are clad in bearded lichens, liverworts, mosses and ferns. In spring, great spotted woodpeckers can be heard drumming on hollow wood, accompanied by the song of wood warbler. Summer visiting birds include tree pipit and nightjar, with woodcock arriving in the autumn. Lesser horseshoe bats enjoy a midnight feast of insects missed by pied flycatchers during daylight hours.

From the car park (Access point), an old gold miner's path contours around the hill, passing through heathland that forms part of the nature reserve. Overall, the route is relatively level but is often rugged, with occasional steep sections. At the top of the reserve, enjoy wonderful views of the Mawddach Valley and the peak of Cadair Idris (893m). The route can be extended to the valley bottom, making an enjoyable but more strenuous circular route.

Above: The Afon Mawddach is visible through the trees of Coed Garth Gell [65] during winter months.

CLYWEDOG GORGE

DOLGELLAU, GWYNEDD MAP REF 66

| Ownership: Private
Designations: IPA, NP, SAC, SSSI
Area: 57ha | Forest type: temperate rainforest
Forest location: SH756185
Explorer Map: OL23 | Ease of access: Moderate
Access point: SH7611318168
///flown.swims.juror
LL40 2RH |

Enjoy stunning views of the Afon Clywedog as it tumbles through its tree-clad gorge. A 4km circular route known as the Torrent Walk, created by engineer Thomas Payne in the 1800s, closely follows both banks of the river. At its far end the trail crosses at Pont Clywedog bridge, near the ruins of an eighteenth-century iron furnace. Along the walk, which includes sections of steps, look for the commemorative bench to plant hunter Mary Richards, who lived at nearby Plas Caerynwch mansion. The sessile oak wood supports hosts of bryophytes, dormouse and lesser horseshoe bat, while otter hunts in the river's swirling pools.

The Afon Clywedog tumbles through Clywedog Gorge [66].

COED SYLFAEN

ABERMAW/BARMOUTH, GWYNEDD MAP REF 67

Ownership: Private (multiple owners) Designations: NP Area: 156ha	Forest type: broadleaved and temperate rainforest Forest location: SH634180 Explorer Map: OL18 or OL23	Ease of access: Moderate Access point: SH6248616602 ///standards.flitting.dairies LL42 1DQ

The steep north-facing slopes overlooking the Mawddach estuary are clad in broadleaved woodlands for several kilometres. The sessile oak, birch and hazel trees growing here host a wonderful diversity of lichens, mosses and ferns, indicating that this is a rare and precious fragment of temperate rainforest. It is surprising that no conservation designations are in place to help protect this ancient wooded area. Most of the land was formerly managed by Sylfaen Farm but now belongs to several private woodland owners. Dry stone walls can be found in many areas under the trees, indicating that much of the hillside was once grazed by livestock.

Exploring the woods requires some effort due to the steep terrain. The Access point is near the Panorama Walk made popular by Victorians, but their tea room and pleasure grounds are long gone. Most visitors will head to the panoramic viewpoint (SH6263816297 | ///flasks.next.cluttered) and back, which is a pleasant short walk. The views across the estuary, south to Cadair Idris and east to the peaks of the Eryri/Snowdonia National Park, are among the best that Britain has to offer.

It is possible to take advantage of a network of public footpaths to the north-east and explore more of the woodland. Keep to the west of Afon Dwynant, a tributary of the Afon Mawddach. Near the top of the hill among a plantation of neighbouring conifers is the standing stone of Cerrig-y-Cledd (SH6428619686 | ///flame.nipping.ounce). Open moorland above the woods provides an enjoyable return option for a spectacular circular walk.

ABERCORRIS

CORRIS, GWYNEDD

MAP REF 71

Ownership: Private: North Wales Wildlife Trust Designations: Area: 1ha	Forest type: broadleaved Forest location: SH749085 Explorer Map: OL23 Ease of access: Moderate	Access point: SH7492608562 ///nods.deeds.unroll SY20 9DB

Hidden among the vast conifer plantations of Dyfi Forest [70], this tiny slither of broadleaved woodland provides valuable habitat for mosses and ferns. Ash and oak with hazel understorey grows on the steep slopes above the Afon Deri. A single footpath crosses the reserve from a small lay-by (Access point) on the unclassified road parallel to the A487. Otherwise, exploring this small wood is quite difficult.

The Mawddach estuary seen from Coed Sylfaen [67].

ARTHOG BOG

FAIRBOURNE, GWYNEDD		MAP REF 68
Ownership: Private: RSPB **Designations:** NP, SAC, SSSI **Area:** 8ha **Forest type:** broadleaved	**Forest location:** SH632139 **Explorer Map:** OL23 **Ease of access:** Easy	**Access point:** SH6287214129 ///bumps.materials.unpainted LL39 1BQ

Carr (wet) woodland of alder and birch surrounds the larger area of bog and mire of this nature reserve near the mouth of the Afon Mawddach. Small birds, including bullfinch, siskin and willow warbler flit between branches, and woodcock skulk along the tree margins. Views across the bog and estuary to the peaks of the Rhinogs are breathtaking. Park at the end of the road next to the tiny railway station of Morfa Mawddach (Access point), and walk 100m back down the Llwybr Mawddach Trail to reach the reserve entrance.

Yellow flag leaves emerge at Arthog Bog [68].

Mosses clad boulders under the canopy of Ystrad Gwyn [69].

YSTRAD GWYN

MINFFORDD, GWYNEDD   MAP REF 69

Ownership: Public: Natural Resources Wales **Designations:** IPA, NNR, NP, SAC, SSSI	**Area:** 17ha **Forest type:** temperate rainforest **Forest location:** SH728117 **Explorer Map:** OL23	**Ease of access:** Moderate/Difficult **Access point:** SH7320911561 ///mash.title.deeds LL36 9AJ

Many walkers attempting the peak of Cadair Idris (893m) will choose the Minffordd path. Although one of the shortest routes, it climbs 788m vertically from the Dôl Idris car park (Access point). The first 132m of the climb passes through the temperate rainforest of Ystrad Gwyn alongside the falls of Afon Cadair. Exploring the remainder of the wood is difficult yet rewarding. The trees support many bryophytes, including Portuguese feathermoss and western earwort. Listen for the *prk-prk* of raven overhead. The scenery is spectacular.

Wild daffodils.

DYFI FOREST

ABERLLEFENNI, GWYNEDD   MAP REF 70

Ownership: Public: Natural Resources Wales **Designations:** NP, UNESCO Dyfi Biosphere Reserve	**Area:** 6,000ha **Forest type:** conifer **Forest location:** SH776097 **Explorer Map:** OL23 or 215	**Ease of access:** Moderate **Access point:** SH7690409316 ///sunblock.submits.matter SY20 9RS

Dyfi forest is an extensive forested area stretching across the counties of Gwynedd and Powys (Central-East). It lies between Dolgellau to the north and Machynlleth to the south. Set among steep and rugged hills, this region was once heart of the Welsh slate mining industry. Most of the forest is coniferous, with broadleaves including oak and beech found nearer valley bottoms and in wooded areas predating the twentieth-century drive to create a strategic reserve of timber. Only a small proportion of its northern extent falls within the Eryri/Snowdonia National Park. It lies at the heart of the UNESCO Dyfi Biosphere Reserve, the only one of its kind in Wales.

There are several places for the visitor to start enjoying the forest. Foel Friog is the most central. From the large car park (Access point), cross the Afon Dulas, passing from Gwynedd into Powys, and climb through the conifers and oaks following a waymarked trail. A viewpoint provides a wonderful view of Foel Crochan and the Aberllefenni slate quarry, and of Godre Fynydd.

Other options include several walking trails among towering Douglas fir and majestic beech at Tan y Coed (SH7550005368 | ///enhancement.many.universal). At the western end of the forest, a choice of waymarked trails start from Nant Gwernol railway station near Abergynolwyn (SH6810206697 | ///royal.smoker.defected) offering superb views and pretty waterfalls en route (arrive by steam train or park in the village).

Looking towards Birds Rock (far right in shadow) from Coed Ysgubor-wen [72].

COED YSGUBOR-WEN

LLANEGRYN, GWYNEDD | MAP REF 72

Ownership: Private: Woodland Trust **Designations:** NP **Area:** 42ha	**Forest type:** broadleaved **Forest location:** SH625068 **Explorer Map:** OL23 **Ease of access:** Moderate	**Access point:** SH6281306745 ///vowel.hotspot.dazzling LL36 9UF

This relatively young broadleaved woodland is surrounded by fragments of temperate rainforest on the northern flanks of the beautiful Dysynni Valley. Situated almost opposite the site and its rapidly growing trees rise the dramatic twin crags of Craig yr Aderyn (Birds Rock). Half a dozen pairs of chough nest on the protected (SSSI) crags, and up to 50–60 have been known to roost on its vertical flanks. Cormorant and raven also nest there. Management of nearby Coed Ysgubor-wen is accounting for the feeding habits of chough, hoping to boost populations by keeping areas clear of bracken. Some areas of the site are quite wet, and host veteran alder trees.

There is a network of permissive paths, some of which involve steep terrain, which also links with a byway (the old 'Milk Path') at the top of the site. The scenery is very fine, particularly views to Craig yr Aderyn and beyond to Cadair Idris at the head of the valley. In the other direction, Twyn and Bae Ceredigion/Cardigan Bay are visible. There is room for a couple of cars to park near the Access point.

Red kite.

In common with the North-West region, the North-East has low forest cover (less than 11 per cent). Incorporating the counties of Denbighshire, Flintshire and Wrexham, this region boasts beautiful coastline and peaceful hills, including the spectacular Clwydian Hills. Sites such as Clwyd Forest [83] Coed Llangwyfan [80] and Coed Nercwys [84] offer wonderful, elevated views across the Vale of Clwyd and beyond. Clocaenog Forest [91] is the largest forest, and its sprawling conifers and wind turbines make it unmissable on the border between the North-East and North-West regions. History abounds, from the ancient earthworks of Offa's Dyke and Wat's

The conifers of Clocaenog Forest [91] and windfarm.

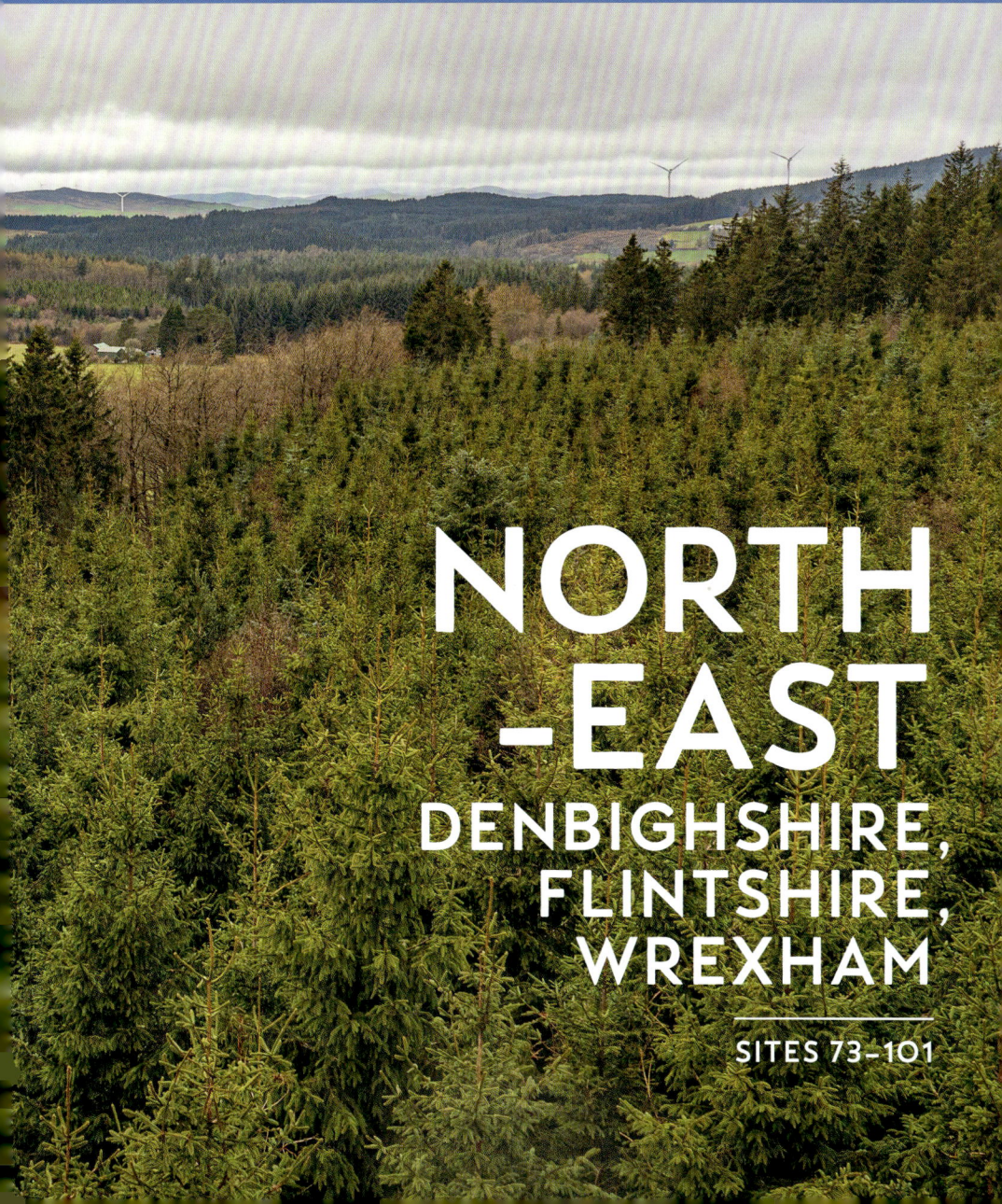

Dyke, which pass through several sites near the border with England, to the huge prehistoric Penycloddiau hill fort at Coed Llangwyfan [80], and the medieval castle and hunting forest at Castell y Waun [98]. Limestone extraction left an indelible mark on the region, as at Coed y Felin [79], Ddol Uchaf [78], Minera Quarry [90] and Pisgah Quarry [96], together with the woods lining former arterial mineral transport routes such as Canal Wood [99]. A formal proposal to create a new fourth national park in Wales is currently underway which would include areas of the Clwydian Range and Dee Valley National Landscape.

NORTH-EAST

DENBIGHSHIRE, FLINTSHIRE, WREXHAM

SITES 73–101

BIG POOL WOOD

GRONANT, FLINTSHIRE   MAP REF 73

Ownership: Private: North Wales Wildlife Trust
Designations: SPA, SSSI
Area: 4ha

Forest type: broadleaved
Forest location: SJ101841
Explorer Map: 265
Ease of access: Easy

Access point: SJ1026783860
///duck.cork.offices
CH8 9NH

Sandwiched between caravan parks and a riding school, and close to the Welsh Channel, this small woodland forms part of a crucially important wildlife corridor for migrating birds. It is situated within the Dee Estuary SPA, which supports huge numbers of waders and waterfowl every winter. The wet woodland, dominated by alder trees, surrounds a large pond. A boardwalk and hide help visitors observe wildlife on the water, in the reedbeds and among the trees. Keep an eye out for spotted flycatcher and treecreeper, and the locally rare giant bellflower. There is room for a few cars on the side of the road close to the Access point.

Treecreeper.

GLAN MORFA

RHYL, DENBIGHSHIRE   MAP REF 74

Ownership: Public: Denbighshire County Council
Designations:
Area: 29ha

Forest type: broadleaved
Forest location: SJ000802
Explorer Map: 264
Ease of access: Easy

Access point: SJ0029880519
///enable.makes.coins
LL18 2AU

Thousands of trees were planted in the mid-2000s on this former landfill site situated between a large housing estate and the estuary of the Afon Clwyd. Its greening has transformed the area, reducing anti-social behaviour, providing valuable green space for local people and improving habitat for wildlife. A network of paths pass through the young trees, including an all-ability trail following the banks of the river. At low tide this is a good place to watch waders feeding on the estuary mudflats.

COED PEN-Y-MAES

TREFFYNNON/HOLYWELL, FLINTSHIRE   MAP REF 76

Ownership: Public: Flintshire County Council
Designations:
Area: 7ha

Forest type: broadleaved
Forest location: SJ194766
Explorer Map: 265
Ease of access: Easy/Moderate

Access point: SJ1939676329
///baths.trusts.reversed
CH8 7BQ

Part of much-loved green space on the outskirts of Treffynnon/Holywell, this long, narrow wood stretches downhill towards the Afon Dyfrdwy/River Dee. Managed by the Friends of Pen y Maes Wood group, the mixed broadleaves provide valuable habitat and shelter for wildlife. Bluebells light up the wood in spring. Enjoy views across to England, including the Wirral and beyond.

COED Y GARREG

CHWITFFORDD/WHITFORD, FLINTSHIRE　　　　　　　　　　　MAP REF 75

Ownership: Private: Mostyn Estate Designations: Area: 21ha Forest type: mixed	Forest location: SJ133782 Explorer Map: 265 Ease of access: Easy/Moderate	Access point: SJ1280778376 ///claw.hypocrite.neon CH8 9DD

In a region rich with prehistoric round barrows and forts, the prominent watchtower on the hill among the trees of Coed y Garreg is shrouded in mystery. Some historians have suggested it was originally a Roman lighthouse, although most believe it is a beacon tower built in the seventeenth century. A stone plaque above its gated entrance, capped by ancient elm beam, marks its restoration in 1897. Its elevated position among the trees provides has, in the past, allowed commanding views north over the Dee estuary, the Wirral and on a clear day even the Cumbrian Hills and the Isle of Man, and in the other direction the Clwydian Hills and Eryri/Snowdonia. However, the broadleaved trees surrounding the tower are rapidly growing, and each year a little more of the view disappears. Perhaps in the future, the owner might once again open up some vistas.

The watchtower at Coed y Garreg [75].

Y GRAIG

TREMEIRCHION, DENBIGHSHIRE   MAP REF 77

Ownership: Private: North Wales Wildlife Trust **Designations:** NL **Area:** 7ha	**Forest type:** broadleaved **Forest location:** SJ084720 **Explorer Map:** 264 **Ease of access:** Easy/Moderate	**Access point:** SJ0838772158 ///progress.grins.bedspread LL17 0UR

Beech, oak and wych elm trees grow on the north side of a limestone outcrop and quarry, whose summit offers wonderful views across the Vale of Clwyd. Some large Scots pine and old pollard beech trees are present. Bluebells, wood sanicle and wood melick thrive under the trees, and common rock-rose in the grassland. At dusk, look for glow-worms along the woodland edge.

The nature reserve is accessed over a stone stile next to a field gate, and the path is steep in places. A small car might be parked carefully next to the entrance, otherwise along the side of the B5429.

Rain runs down the stems of an ancient beech tree at Y Graig [77].

DDOL UCHAF

AFON-WEN/AFONWEN, FLINTSHIRE   MAP REF 78

Ownership: Private: North Wales Wildlife Trust
Designations: SSSI
Area: 4ha

Forest type: broadleaved
Forest location: SJ141713
Explorer Map: 265
Ease of access: Easy/Moderate

Access point: SJ1429071382
///orbit.loudly.remotest
CH7 5UN

Tufa limestone was once quarried on this site, but the wet clay hollows left behind now host all three species of native newts (great-crested, palmate and smooth). Ash, sycamore and willow form the main canopy, while below, elder, hawthorn and especially hazel provide valuable habitat for a healthy population of resident dormice. Patches of trees are routinely thinned to allow sunlight in, providing warmth for basking grass snakes and stimulating the regeneration of woodland flora. A circular path can be followed around the woodland, which is sometimes muddy.

Hawthorn leaves.

COED Y FELIN

HENDRE, FLINTSHIRE   MAP REF 79

Ownership: Private: North Wales Wildlife Trust
Designations:
Area: 10ha

Forest type: broadleaved
Forest location: SJ190677
Explorer Map: 265
Ease of access: Easy/Moderate

Access point: SJ1954667785
///subsystem.regulates.backyards
CH7 5QQ

The limestone extracted in the large working quarry just north of this nature reserve also creates the ideal soil conditions for one of Britain's rarest flowering plants, the Deptford pink, found at only four sites in Wales (and 30 in England). Next to the hay meadow where it thrives lies an ancient semi-natural woodland, which once provided structural timbers for a local coal mine and firewood for its workers and families. Look for wood anemone under the trees, and pied flycatchers feeding from their favourite branches among the ash and sycamore. An old railway line can be easily followed next to a strip of wet woodland. Other paths climb the hill, which are occasionally steep with some steps. Not to be confused with a wood of the same name [278] in the South-Central region.

Even on a misty day, Coed Llangwyfan [80] is a wonderful site to explore.

COED LLANGWYFAN

LLANGWYFAN, DENBIGHSHIRE — MAP REF 80

Ownership: Public: Natural Resources Wales **Designations:** NL **Area:** 51ha	**Forest type:** mixed **Forest location:** SJ134666 **Explorer Map:** 265 **Ease of access:** Moderate	**Access point:** SJ1389066855 ///inquest.impose.tissue LL16 4NB

This large plantation of pine and other conifers grows on the steep valley sides below one of Britain's largest prehistoric monuments. Penycloddiau hill fort covers an impressive area (21ha), its multiple ramparts of banks and ditches occupying a prominent ridge. These were once topped with a wooden palisade, providing added protection for the many roundhouses and other buildings inside. Another hill fort, Moel Arthur, lies nearby to the south-east.

Follow a 3km-long circular walk through the forest, which is waymarked, to enjoy scenery visible between the mature pine and spruce, and to higher slopes where oak and birch grow. Alternatively, from the car park (Access point) head west, following the edge of the trees and Offa's Dyke Path, to climb up Penycloddiau (440m), where spectacular views can be enjoyed across the forest and the Clwydian Hills.

88 The Forest Guide: Wales

COED PWLL-Y-BLAWD

CADOLE, FLINTSHIRE   MAP REF 81

Ownership: Public: Denbighshire Country Park **Designations:** NL, SAC, SSSI **Area:** 100ha	**Forest type:** broadleaved **Forest location:** SJ194630 **Explorer Map:** 265 **Ease of access:** Easy/Moderate	**Access point:** SJ1972162569 ///stumps.clashes.bedspread CH7 5SE

Situated with the popular Loggerheads Country Park, this ribbon of woodland follows the narrow valley of the Afon Alun/River Alyn through the dramatic karst landscape, which once inspired British landscape artist Richard Wilson (1714–1782). The valley deepens into a gorge upstream, with steep limestone sides, while sinkholes mean that in dry periods the riverbed dries up as the water disappears into channels deep underground. A footpath runs through the site known as the Leete Path, following an artificial water channel (a leat), which once carried water to mines and later to Pentre Mill, a waterwheel used to mill corn and to drive a circular saw to cut timber. Among the ash, oak and wych elm trees, look for pied flycatcher, adder, wood anemone and ramsons. Dipper and kingfisher feed in the river, while areas of limestone grassland are rich with butterflies, notably green hairstreak and grizzled skipper. Several waymarked routes can be followed, the Leete Path being the most level, while others can involve sections with steps, especially to a viewpoint offering scenic views of Moel Famau, and Clwyd Forest [83]. Devil's Gorge is reached about 2km upstream, crossed by a pedestrian bridge near to crags popular with climbers. Parking charge.

Looking towards Clwyd Forest [83] and the peak of Moel Famau from Coed Pwll-y-Blawd [81].

ABERDUNA

MAESHAFN, GWYNEDD

MAP REF 82

Ownership: Private: North Wales Wildlife Trust **Designations:** NL **Area:** 20ha	**Forest type:** broadleaved **Forest location:** SJ205615 **Explorer Map:** 265 **Ease of access:** Moderate	**Access point:** SJ2054761663 ///shares.recap.nappy CH7 5LD

Herb-Paris, wood sorrel and goldilocks buttercup thrive under the ash, oak, sycamore and wild cherry trees of this attractive nature reserve. Birch, elder, hazel, holly and several willow species add diversity to woodland edges and the understory. Look for common redstart and willow warbler during summer months. Adjacent to the woodland, the site's shallow soils overlying limestone result in a very rich grassland, with a wide range of wildflowers including wild thyme and common rock rose, and the yellow club fungus. This is a good site for woodland and grassland butterflies, including small pearl-bordered and brown argus.

A number of paths can be followed around the hilly reserve, circumnavigating the large quarry that cuts into the site, and these can be steep in places. On a clear day the views are very attractive, including Clwyd Forest [83] and Moel Famau (554m), the highest peak of the Clwydian Range. The Access point is the car park of the North Wales Wildlife Trust's offices, which are only open during normal working hours.

CLWYD FOREST

LLANFERRES, DENBIGHSHIRE

MAP REF 83

Ownership: Public: Natural Resources Wales	**Forest type:** conifer	**Access point:** SJ1714361065
Designations: NL	**Forest location:** SJ167617	///effort.height.dupe
Area: 288ha	**Explorer Map:** 265	CH7 5SH
	Ease of access: Moderate/Difficult	

Most people pass through this large plantation forest en route to Moel Famau (554m), the highest peak of the Clwydian Range. Sitka spruce currently dominates the forest but it is being diversified. The forest rides are easy to follow but climb quite steeply. Open ground is reached after 1.5km, and the summit lies another 600m away. If the summit is your destination, take care to dress appropriately for mountain weather. The forest shelters in the eastern lee of the hill, but along its ridge isolated lodgepole pine trees exhibit krummholz form (stunted and wind-bent). Once out of the trees, the views over Clwyd Forest and the Vale of Clwyd are spectacular, while the wind turbines of Clocaenog Forest [91] are visible in the far distance. The Access point given is the lower car park, which has an entry barrier (charge), offering toilets and a children's play area. Starting at the higher car park (charge) and following the Offa's Dyke Path is a better option for those more interested in the scenery.

The upper limits of Clwyd Forest [83].

COED NERCWYS

MAESHAFN, FLINTSHIRE

MAP REF 84

Ownership: Public: Natural Resources Wales
Designations: NL, SM
Area: 167ha

Forest type: coniferous
Forest location: SJ219585
Explorer Map: 265
Ease of access: Easy/Moderate

Access point: SJ2183159269
///powers.dude.mastering
CH7 4DD

This large plantation forest, created in the 1960s, is popular with walkers and cyclists. A number of waymarked routes can be followed, which vary in length and difficulty. Climb to the triangulation point to enjoy excellent views across to the Clwydian mountains. Within the forest lies a prehistoric cairn circle dating to the Bronze Age (SJ2210058044 | ///operating.download.dynamic). Llyn Ochin – a remnant of old lead mining activities – is now a bog, providing habitat for amphibians and water-loving plants.

MARFORD QUARRY

MARFORD, WREXHAM

MAP REF 85

Ownership: Private: North Wales Wildlife Trust
Designations: SSSI
Area: 11ha

Forest type: broadleaved
Forest location: SJ357560
Explorer Map: 256
Ease of access: Easy/Moderate

Access point: SJ3572156337
///sunk.lower.fists
LL12 8TF

The glacial moraine at Marford was quarried for more than 40 years, most famously to provide aggregate for the Mersey Tunnel, opened in 1934. After it closed in 1971 the main quarry was designated a nature reserve and is now owned by the local wildlife trust, while nearby Maes y Pant [86], which also formed part of the site, is owned by the local community.

The deep quarry walls form a natural amphitheatre, and at its centre a lone silver birch grows on a small mound. The quarry bowl is criss-crossed by paths, which are easy to explore. The rim of the quarry walls can also be ascended, providing good views over the Dee Valley.

The reserve is especially rich in invertebrate life, which take advantage of the sandy soil, particularly bees, wasps and ants, the latter of which provide food for green woodpeckers. Bee and pyramidal orchids are among the many flowers found in open glades and surrounding grassland.

Early purple orchid.

MAES Y PANT

GRESFORD, WREXHAM

MAP REF 86

Ownership: Private: Maes y Pant Action Group Designations: Area: 25ha	Forest type: mixed Forest location: SJ354555 Explorer Map: 256 Ease of access: Easy	Access point: SJ3540655173 ///drooling.senior.communal LL12 8RF

This community-owned woodland is loved by local people. Since its purchase in 2011, 'MyPag' volunteers have created a network of paths (including 3km of surfaced trails) with interpretation boards, undertaken conservation work, and installed sculptures. The 28,000 Corsican pine trees originally planted on the former quarry site are being gradually felled and replaced with a more diverse range of tree species. Look out for sparrowhawk and goldcrest, and listen for the hoot of tawny owl at dusk. The flora is especially diverse, with more than 200 species recorded, including bee orchid, harebell and broad-leaved helleborine. Among many interesting moths, large emerald and buff-tip have been found on the site. A viewpoint reached by ascending 109 steps offers a great view across the Dee Valley, and on a clear day the city of Liverpool in the far distance. The site is immediately adjacent to Marford Quarry [85], an SSSI owned by the North Wales Wildlife Trust, which can be reached from the same Access point.

COED CILYGROESLWYD

PWLL-GLAS, DENBIGHSHIRE

MAP REF 87

Ownership: Private: North Wales Wildlife Trust Designations: SSSI Area: 4ha	Forest type: mixed Forest location: SJ124553 Explorer Map: 256, 264 or 265 Ease of access: Moderate	Access point: SJ1273755331 ///school.soonest.searches LL15 2EQ

An unusual mix of ash, oak and yew form the main canopy of this attractive woodland. Seeds within the sticky bright-red yew arils are a favourite food for the hawfinch, while the site's rich invertebrate life feeds pied flycatchers. Despite the ground flora having several ancient woodland indicator species, including sanicle and sweet woodruff, this is not a particularly old woodland. Gaps in the underlying limestone pavement are thought to have conserved some flora from the ancient woodland that was here before being cleared and grazed. This is one of only two sites in Wales where the county flower of Denbighshire, limestone woundwort, is found. Look for its pretty pink flowers among the hedge bottoms and path verges.

The site can be reached by walking up a country lane from Pwll-glas, although the Access point given is closer by. Park in a small lay-by on the east side of the bridge over the Afon Clwyd and walk to the entrance of the reserve directly opposite, taking care when crossing the A494. Paths are steep in places and slippery when wet.

GRAIG WYLLT

GRAIG-FECHAN, DENBIGHSHIRE   MAP REF 88

Ownership: Private: North Wales Wildlife Trust **Designations:** NL, SM **Area:** 4ha	**Forest type:** broadleaved **Forest location:** SJ148552 **Explorer Map:** 256 **Ease of access:** Moderate/Difficult	**Access point:** SJ1476154395 ///combining.issuer.crouches LL15 2EU

Appropriately named 'wild rock', this craggy outcrop is partially clad in ancient woodland. Its ash and oak trees support a rich ground flora of bluebell, wild garlic and wood anemone. Grassland areas are graced by grayling, purple hairstreak and ringlet butterflies, and the distinctive white-green flowers of the greater-butterfly orchid. Special care is advised near the top of the steep-sided crags, but the views across the Vale of Clwyd are truly wonderful. On the opposite side of the valley, in Coed Henblas, is the site of a large medieval moated homestead (SJ1471555185 | ///defended.majors.quieter) which can be explored via a public footpath. The reserve entrance is reached by walking north for 600m from the Access point. Car parking is available for customers of the pub, 100m north of the Access point. Dogs are not permitted on the reserve.

COED LLANDEGLA

PEN-Y-STRYT, WREXHAM   MAP REF 89

Ownership: Private: Church Commissioners for England **Designations:** NL **Area:** 583ha	**Forest type:** conifer **Forest location:** SJ220506 **Explorer Map:** 256 **Ease of access:** Moderate	**Access point:** SJ2390652371 ///handrail.plates.sandbags LL11 5UL

The large coniferous forest originally planted in the 1970s is now owned by the Church Commissioners for England, an organisation that is generally rather shy about the many tens of thousands of hectares of forest that it has invested in across Britain and overseas. Some have criticised the owner for focussing more on its investment portfolio than the delivery of social and environmental benefits across its many forest sites, yet here it is exemplary. At Coed Llandegla, Sitka spruce dominates the forest, with some minor areas of other conifer species and 5 per cent native broadleaves. Continuous cover forestry is being introduced to help create a more diverse and resilient forest. Some conifers are being retained beyond their usual economic age to help restructure the forest. The site is popular with hardcore mountain bikers, with four graded trails offered, while there are several walking trails, one of which circumnavigates Pendinas Reservoir, all of which start at a privately run visitor centre at the east end of the site (Access point). The Offa's Dyke Path passes through the forest at its west end, providing an alternative and more peaceful access option (SJ2074551191 | ///showdown.seasonal.hockey), the route climbing quite steeply up through the forest to reach open ground, where the owners are promoting habitat for black grouse.

Hawthorn trees laden with haws (fruits) at Minera Quarry [90].

MINERA QUARRY

MINERA, WREXHAM   MAP REF 90

Ownership: Private: North Wales Wildlife Trust **Designations:** NL, SM, SSSI **Area:** 19ha	**Forest type:** broadleaved **Forest location:** SJ251521 **Explorer Map:** 256 **Ease of access:** Easy/Moderate	**Access point:** SJ2582851934 ///drama.confusion.former LL11 3DE

This large quarry produced valuable limestone until it closed in 1994, and for hundreds of years before that the site was a lead mine. Today, nature is slowly recolonising the site, with woodland thriving on older abandoned parts and lime-rich grassland dominating other areas. The crags are used by nesting peregrine falcon and raven, while buzzard and tawny owl reside within the woodland. Other woodland birds include blackcap, common redstart and spotted flycatcher. The grassland supports many rare flowering plants and invertebrates, notably belted clearwing, grayling and mountain bumblebee moths. Look closely at the boulders near the entrance, where fossils are easy to spot, and along the edge of the stream a row of lime kilns can still be seen. The site is quite easy to explore, although the paths often consist of loose gravel. There are very fine views from the higher points of the reserve.

CLOCAENOG FOREST

CLOCAENOG, DENBIGHSHIRE   MAP REF 91

Ownership: Public: Natural Resources Wales **Designations:** SM **Area:** 1,517ha	**Forest type:** conifer **Forest location:** SJ019530 **Explorer Map:** 264 **Ease of access:** Easy/Moderate	**Access point:** SJ0366151165 ///emperor.gossiping.youngest LL15 2DT

This very large plantation forest straddles two regions, spreading into the North-West region and the county of Conwy. Its first conifers were planted by owner Lord William Bagot in the 1830s and later felled to support the First World War (pit props and trench timbers) and then replanted by the Forestry Commission. It remains a working forest, with areas of felling and replanting throughout. Notable wildlife includes red squirrel and goshawk, while black grouse hold their dramatic courtship leks in remote forest clearings.

The Access point at Bod Petryal includes a picnic site and the beginning of a cycle trail. Prehistoric stone circles and the homestead of Cefn Banog Ancient Village (SJ0184951021 | ///yachting.painters.boast) are also both within reach. It is one of four car parks along the B5105 alone, with others elsewhere around the periphery of the forest. A wide range of waymarked trails criss-cross the forest, offering great recreational opportunities for walkers and cyclists. The Pincyn Llys monument, erected to commemorate the original planting, stands in a clearing at the highest point (502m), offering expansive views across the forest.

Opposite: An avenue of beech near Bod Petryal in Clocaenog Forest [91].

Below: Clocaenog Forest [91] and windfarm seen from Moel Famau.

LEWIS WOOD

WRECSAM/WREXHAM, WREXHAM   MAP REF 92

| Ownership: Private: National Trust
Designations: HPG
Area: 16ha
Forest type: broadleaved | Forest location: SJ339487
Explorer Map: 256
Ease of access: Easy/Moderate | Access point: SJ3356548614
///latter.reclaimed.cadet
LL13 7EZ |

Forming a peri-urban fringe around the south of Wrecsam/Wrexham, this narrow wood along the banks of the Afon Clywedog is a wonderful asset for local people and a valuable habitat for wildlife. The wood is about 300 years old and hosts carpets of bluebells in spring. It is part of the Erddig Hall estate, which includes several other areas of woodland, all of which can be linked by following public footpaths to make a longer circular route. The Clywedog Trail also passes through the trees. The entrance to the wood is a short distance north from the car park at Sontley Bridge (Access point), while the attractive parkland of Erddig Country Park is accessible immediately to the west.

Bluebell flowers at Lewis Wood [92].

WURTHYMP WOOD

WORTHENBURY, WREXHAM   MAP REF 93

| Ownership: Private
Designations:
Area: 7ha
Forest type: broadleaved | Forest location: SJ425461
Explorer Map: 257
Ease of access: Easy | Access point: SJ4208446095
hissing.degrading.blanking
LL13 0AR |

A young broadleaved plantation created in 2006, this wildlife-friendly woodland is being managed using traditional coppice silviculture to produce charcoal and other sustainable woodland products. The owner was recently commended for its management by the Royal Forestry Society. The Access point is a public footpath at the junction of Frog Lane and Mulsford Lane. Walk across two small fields to enter the west end of the wood.

TREVOR HALL WOOD

TREVOR, DENBIGHSHIRE
MAP REF 94

Ownership: Private: Trevor Hall Estate Designations: HPG, NL Area: 61ha	Forest type: mixed Forest location: SJ250424 Explorer Map: 255 Ease of access: Moderate	Access point: SJ2588142341 ///quilting.crisis.scooter SJ258423

Growing on the steep south-facing slopes above Trevor Hall, the trees are steeped in history and cloaked in a little mystery. While the hall remains out of sight below, this is a peaceful wood situated in the Clwydian Range and Dee Valley NL. Once a magnificent country mansion built in 1742, the hall fell to utter ruin and was even used to shelter cattle before a complete renovation was undertaken in the late 1990s. It is now a family home and a venue hired out for large functions. The visitor can follow several different paths through the trees, some of which are overgrown, though the Offa's Dyke Path, which passes through the wood, is clearly waymarked. On its northern perimeter is Pen y Gaer Hillfort, while the explorer continuing beyond the trees to the west will reach the aptly named Panorama Walk and beyond the prominent Iron Age hill fort of Castell Dinas Brân. Once clear of the trees, there are wonderful views across the valley, taking in the meanders of the Afon Dyfrdwy/River Dee and the Llangollen Canal. Access the wood from the lane near Trevor Hall Chapel (but do not park here during church services) and follow the Offa's Dyke Path into the wood.

PEN Y COED

LLANGOLLEN, DENBIGHSHIRE
MAP REF 95

Ownership: Private: Woodland Trust Designations: NL Area: 28ha	Forest type: mixed Forest location: SJ225416 Explorer Map: 255 Ease of access: Moderate	Access point: SJ2200241776 ///galloping.cultivation.foil LL20 8LR

Lying just east of Llangollen, the trees of Pen y Coed cover a prominent long hill overlooking the town. The tree-lined ridge is also visible from across the Dee Valley, including from Trevor Hall Wood [94]. This is ancient semi-natural woodland, although many of its old trees were felled and replaced with conifers in the mid-nineteenth century, but management is slowly transitioning it back into a broadleaved wood. The gentler slopes to the south of the hill shows signs of old coppice management, once being part of the land owned by the Cistercian abbey of Valle Crucis, whose magnificent ruins lie 3km as the crow flies to the north-west. A good diversity of woodland wildlife can be seen throughout the year, while carpets of bluebells and ramsons light up the wood in spring. A public footpath follows an old track below the wood's very steep north-facing slopes, allowing wonderful glimpses to the Afon Dyfrdwy/River Dee and beyond. Two waymarked trails are also provided. All routes require some ascending, and can be steep in places. The wood is best accessed from town (Brook Street), where there is ample parking.

PISGAH QUARRY

FRONCYSYLLTE, WREXHAM — MAP REF 96

Ownership: Private: North Wales Wildlife Trust **Designations:** NL **Area:** 1ha	**Forest type:** broadleaved **Forest location:** SJ268411 **Explorer Map:** 255 **Ease of access:** Easy/Moderate	**Access point:** SJ2681441087 ///paints.beyond.armrests LL20 7SF

Situated in the Clwydian Range and Dee Valley NL, offering wonderful views across the Vale of Llangollen, including the famous Pontcysyllte Aqueduct on one side and Castell Dinas on the other, this small woodland and nature reserve is a little gem. Its ash, sycamore and wych elm trees thrive on the site of a former quarry where limestone was once extracted to be converted in local kilns to quicklime. Used as cement for buildings and fertiliser for fields, it was sent far and wide via the canal network. An open glade in the wood is full of flowering plants, which thrive on the limestone soil. Walking near the small car park (Access point) is easy, but beyond soon becomes quite steep.

Castell Dinas seen from Pisgah Quarry [96].

KNOLTON WOOD

OWRTYN/OVERTON, WREXHAM   MAP REF 97

Ownership: Private **Designations:** **Area:** 13ha **Forest type:** mixed	**Forest location:** SJ348407 **Explorer Map:** 256 **Ease of access:** Moderate	**Access point:** SJ3541140351 ///hiked.strutting.small SY12 9EW

South of Wrecsam/Wrexham, the border between England and Wales is tortuous. It follows the Afon Dyfrdwy/River Dee for some of its length, but heads south along the Shell Brook through the trees of Knolton Wood. This mixed woodland on the east side of the brook lies within Wales and covers 13ha. A portion of the 64km-long Wat's Dyke Way runs through the wood and is clearly waymarked. Wat's Dyke is an ancient linear earthwork consisting of a bank and ditch that runs broadly parallel and sometimes very close to the better-known Offa's Dyke. A pretty waterfall can be found near the remains of Knolton Mill, surrounded by wych elm trees. Look for dipper, grey wagtail and kingfisher. The Afon Dyfrdwy/River Dee forms the northern boundary of this crescent-shaped wood. There was once a hand-operated winch next to the Boat Inn to carry pedestrians across the river, whose pools and riffles are now popular among fly-fishers. An enjoyable circular walk can be completed by continuing to Llan-y-cefn Wood.

CASTELL Y WAUN (CHIRK CASTLE)

Y WAUN/CHIRK, WREXHAM   MAP REF 98

Ownership: Private: National Trust **Designations:** HPG, NL, SM, SSSI **Area:** 30ha **Forest type:** mixed	**Forest location:** SJ271377 **Explorer Map:** 240 **Ease of access:** Easy/Moderate	**Access point:** SJ2673138342 ///outraged.dentistry.carefully LL14 5BL

From the medieval period, a vast hunting forest and deer park extended across many thousands of hectares of Welsh countryside west of Y Waun/Chirk. Today, the estate is considerably smaller, yet provides wonderful habitat for wildlife and a rewarding visitor experience. A prominent section of the defensive earthwork of Offa's Dyke passes within 200m of the castle. The parkland hosts some 650 veteran trees (two-thirds of which are oak), providing habitat for mosses, lichens, ferns and fungi. Treecreeper and great spotted woodpecker thrive on a rich diet of insects found among the trees' decaying wood. A circular waymarked woodland trail (1.6km) passes through Baddy's Wood (broadleaved) and Deer Park Wood (coniferous), offering great views and a chance to watch birds from a hide. Open only from March until October. Entry charged.

Fronds of a male fern.

COED COLLFRYN

LLWYNMAWR, WREXHAM   MAP REF 100

Ownership: Private: Woodland Trust
Designations:
Area: 7ha

Forest type: broadleaved
Forest location: SJ218370
Explorer Map: 240
Ease of access: Moderate

Access point: SJ2207237168
///sorry.paves.creatures
LL20 7AY

Rising up the steep valley sides south of the Afon Ceiriog, Coed Collfryn is an attractive small woodland. Its canopy is mostly sessile oak, although many of them are hybridised with pedunculate oak, accompanied by silver birch. Within its lower margins a more diverse mixture of trees is found, including ash, sycamore and wild cherry, with holly and rowan. Keep a look out for pignut, a dainty umbelliferous plant whose edible tubers (swollen roots that can grow 15–20cm long) taste like hazelnut.

Honeysuckle.

Honeysuckle twines around many trees, providing nectar for invertebrates. Public footpaths criss-cross the site and are steep in places. There is room for several cars to park 50m north of a kissing gate (Access point), which leads to the wood.

FENN'S WOOD

BETTISFIELD, WREXHAM   MAP REF 101

Ownership: Public: Natural Resources Wales
Designations: IPA, NNR, SAC, SSSI
Area: 57ha

Forest type: mixed
Forest location: SJ499381
Explorer Map: 241
Ease of access: Moderate

Access point: SJ4932835410
///sectors.curvy.fattening
SY13 2RX

The sphagnum moss and peat of the raised bog at Fenn's, Whixall and Bettisfield Mosses National Nature Reserve lies 8m deep in places, fed only by rainwater. 'The Marches Mosses' as they are known locally, straddle England and Wales. Fenn's Wood borders the eastern and northern margins of the raised bog, providing valuable cover and alternative habitat for wildlife. Listen for yellowhammer singing from the tops of small trees, intermingling with the calls of curlew and lapwing over the bog. Long-eared owl and nightjar hunt along the woodland edges. From the car park near the Llangollen Canal (Access point, which lies in England), walk west towards Moss Cottages, before heading north, crossing into Wales to reach the wood. The paths are often very wet underfoot.

CANAL WOOD

Y WAUN/CHIRK, WREXHAM

MAP REF 99

Ownership: Private: Canal and River Trust **Designations:** UNESCO World Heritage Site	**Area:** 13ha **Forest type:** broadleaved **Forest location:** SJ282382 **Explorer Map:** 240	**Ease of access:** Easy **Access point:** SJ2841137782 ///slowly.kicked.geologist LL14 5AA

Amble alongside narrowboats on the Llangollen Branch of the Shropshire Union Canal as they pass through a cutting dissecting this beautiful wood of lime, sycamore and ash. Look too for several majestic ancient sweet chestnut trees on the top of the bank opposite the industrial estate. The towpath is flat and surfaced, suitable for access on foot or wheels, forming part of Sustrans Route 84. The entrance to the wood is via the towpath at its southern end (Access point), which is easily reached from town. The adventurous, armed with a torch, can also head south after dropping down the short ramp to reach the towpath, and walk through the 421m-long Chirk Tunnel known as 'The Darkie'. The tunnel emerges at the stunning 220m-long Chirk Aqueduct, completed in 1801 by William Jessop and Thomas Telford, which carries the canal 21m above the Afon Ceiriog. The views across the partially wooded valley are spectacular, with England lying on the opposite bank after a short and very dramatic walk along the aqueduct towpath.

A narrowboat moored under sweet chestnut trees in Canal Wood [99] near the east entrance to the Chirk Tunnel.

Often defined as part of a larger region termed Mid Wales under modern political boundaries, the western area is defined by the county of Ceredigion and termed Central-West in this guide. It has the highest forest cover (almost 18 per cent) of any region, while many of its sites are rich with bryophytes (see pp.116–17) and considered precious fragments of temperate rainforest, including Coed Cwm Clettwr [105], Coed Cwm Einion [104], Coed Simdde Lwyd [115] and Coed Rheidol [113]. Birdlife is wonderfully diverse, including the iconic woodland species of lesser spotted woodpecker,

The Afon Ystwyth at Hafod [120].

pied flycatcher, redstart and wood warbler, which can be seen at Coed Cwm Clettwr, Coed Penglanowen [116] and Coed Simdde Lwyd among others. Innovative conservation work is underway in the region, as demonstrated by Coetir Anian/Cambrian Wildwood at Bwlch Corog [102]. The largest forest in this guide is impossible to miss, with the 6,000ha of Tywi Forest [127] covering a vast area yet attracting relatively few visitors, unlike Bwlch Nant yr Arian Forest [112], which is a popular destination for mountain bikers and a popular place to watch red kites.

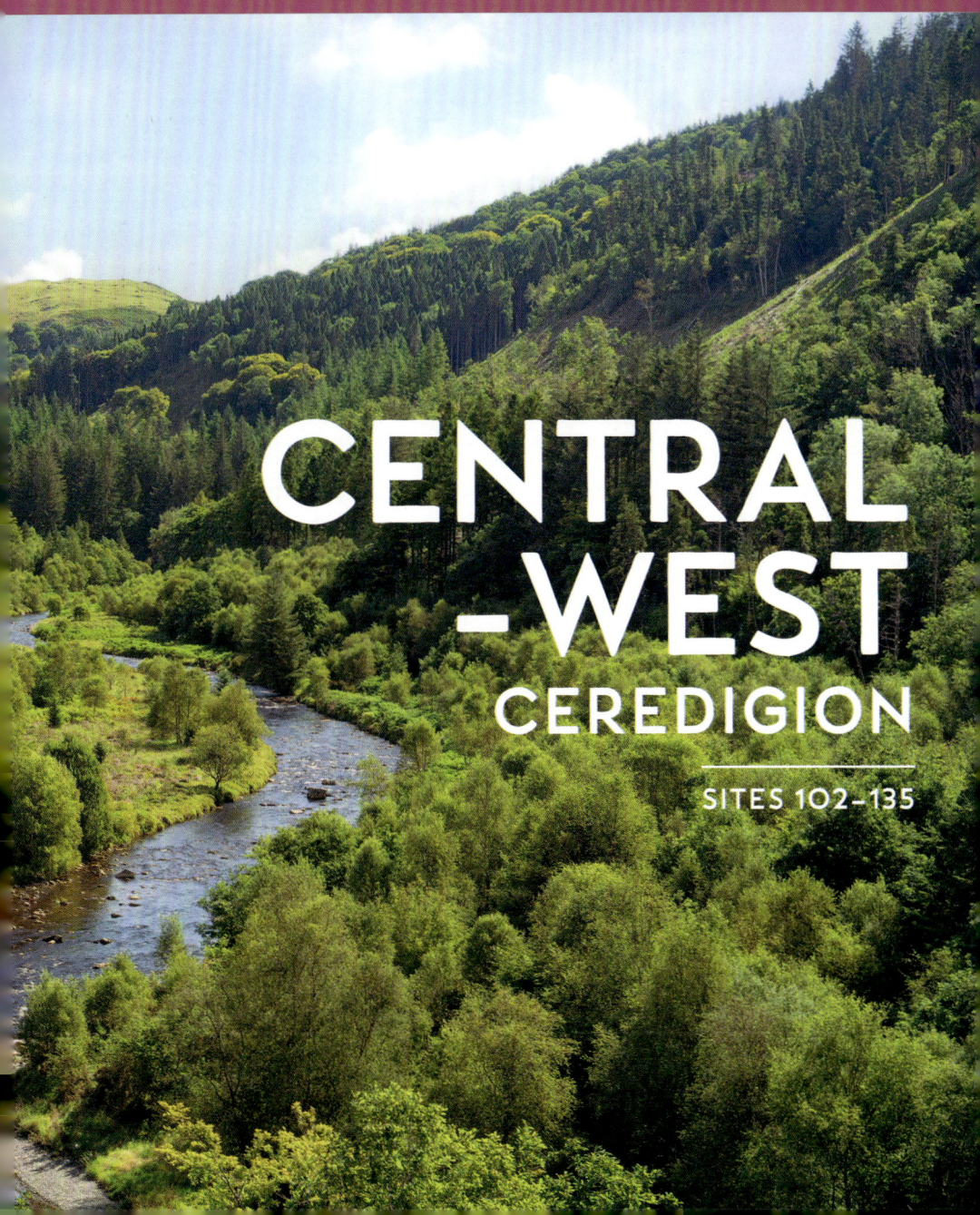

CENTRAL -WEST
CEREDIGION
SITES 102–135

BWLCH COROG

GLASPWLL, CEREDIGION — MAP REF 102

Ownership: Private: Coetir Anian **Designations:** IPA **Area:** 5ha **Forest type:** broadleaved	**Forest location:** SN738964 **Explorer Map:** OL23 **Ease of access:** Moderate	**Access point:** SN7396596917 ///operation.cookbooks.magnitude SY20 8UA

Coetir Anian/Cambrian Wildwood is working to restore natural ecosystems in Mid Wales. A small massif in north Ceredigion was its first land acquisition (140ha from freeholder Woodland Trust), lying next to the large conifer plantation of Maesycilyn. Bwlch Corog means 'pass of the wooded precipice by the river'. Only small areas of woodland are currently present (some of which fall within the Cambrian Mountains Woodlands IPA), but the charity is actively assisting natural tree regeneration. Grazing has been introduced and drains blocked to enhance wildlife habitats, and thousands of trees planted. The charity's ambition is for woodland to extend naturally up the slopes, gradually transitioning to bog and open moorland. In future, perhaps beaver, pine marten, wild boar and wild cat will join the Highland cattle and wild horses on the site.

Near the village of Glaspwll, look for a sign next to the small bridge over the Afon Llyfnant (SN7392397516 | ///coolest.fluffed.adjusted) and follow the track that climbs up the valley of the Nant y Factory. The Access point is the northern tip of the site boundary, which extends southwards to the summit of Bwlch Corog (388m).

COED CWM EINION

FURNACE, CEREDIGION — MAP REF 104

Ownership: Public: Natural Resources Wales **Designations:** IPA, SAC, SSSI, UNESCO Dyfi Biosphere Reserve	**Area:** 21ha **Forest type:** temperate rainforest **Forest location:** SN685948 **Explorer Map:** OL23	**Ease of access:** Difficult **Access point:** SN6844195300 ///tactical.nicely.enacted SY20 8ND

Dominated by small-leaved lime and ash, an unusual mixture so far westward, this semi-natural woodland is also a rare fragment of temperate rainforest. It is one of the three core areas of the UNESCO Dyfi Biosphere Reserve (see also p.80 and p.109). The main body of the wood lies south of the Afon Einion, where multitudes of bryophytes and ferns grow in the shelter of the steep-sided gorge. Among these are many rare or scarce species, such as hay-scented buckler fern, Tunbridge filmy-fern, the lobed lichen *Parmelia robusta*, and the liverwort western featherwort.

Exploring the gorge is not easy. Public footpaths cross the river at three different points, but otherwise it is difficult to reach. Park in the village in a public car park (Access point) and navigate the mix of unclassified roads and public footpaths.

Small-leaved lime.

YNYS-HIR

EGLWYS-FACH, CEREDIGION

MAP REF 103

Ownership: Private: RSPB
Designations: SSSI, UNESCO Dyfi Biosphere Reserve
Area: 38ha

Forest type: broadleaved
Forest location: SN682962
Explorer Map: OL23
Ease of access: Easy/Moderate

Access point: SN6825496251
///firepower.boxing.silks
SY20 8TA

Ynys-hir is a spectacular nature reserve. Its 800ha embraces a uniquely diverse range of habitats, including estuary mudflats, salt marshes, freshwater reedbeds, peatbog, broadleaved woodland, heath and moorland. The breadth of wildlife to be seen is therefore incredibly varied, and it is quite possible to see 60 bird species or more during a single visit. In the woodland alone, pied and spotted flycatcher, redstart and wood warbler are notable attractions, especially while they sing to mark their territories above carpets of bluebell and wood anemone in the spring. Osprey, which breeds on the nearby Cors Dyfi reserve, can often be seen over the water in summer.

Spotted flycatcher hunting.

Nightjar fly in the margins of day and night where heath meets woodland, while lesser spotted woodpecker are more commonly seen among the trees. Meanwhile, huge numbers of geese and waders feed in the Dyfi/Dovey estuary, and otter and kingfisher in the quiet freshwater pools. Several trails can be followed, including a 2.5km waymarked woodland trail, which includes the panoramic treetop Ynys-hir hide, offering spectacular views across the valley. This route passes by many ancient oak and standing dead trees, which provide wonderful habitat for invertebrates, making this the best route to spot the rare woodland birds that feed on them.

The view from the treetop wildlife hide at Ynys-hir [103].

COED CWM CLETTWR

TRE'R-DDÔL, CEREDIGION MAP REF 105

Ownership: Private: Wildlife Trust of South and West Wales **Designations:** SSSI **Area:** 16ha	**Forest type:** temperate rainforest **Forest location:** SN666920 **Explorer Map:** OL23 **Ease of access:** Moderate	**Access point:** SN6669692231 ///nursery.breezy.instructs SY20 8PR

Where the Afon Clettwr tumbles off the Mynyddoedd Cambrian/Cambrian Mountains on its way to the Dyfi/Dovey estuary, its gorge is lined with woodland. The SSSI covers the south side of the river, where sessile oak dominates, with occasional pure ash. Dormice live here, and the site is rich with bryophytes and ferns, including polypody, beech fern and oak fern. Sweet woodruff and yellow archangel can be found in some areas, especially where ash thrives. Dipper and grey wagtail feed in the river, and nuthatch, pied flycatcher, redstart and wood warbler among the trees.

On the north side of the river, felling of a former 20ha plantation of western hemlock was completed in 2000, and its slopes are now regenerating from heath to woodland. Young silver birch trees are leading the transition.

There is a small lay-by near the reserve entrance. Paths are generally steep and uneven.

GLANMOR FACH

ABERYSTWYTH, CEREDIGION MAP REF 107

Ownership: Private **Designations:** **Area:** 11ha **Forest type:** conifer	**Forest location:** SN589835 **Explorer Map:** 213 **Ease of access:** Moderate	**Access point:** SN5867184052 ///pocketed.delusions.casually SY23 3DL

Climb the zigzag path up Constitution Hill at the north end of Aberystwyth sea front – or take the cliff railway – to reach its summit and continue another 1km along the Ceredigion Coast Path to reach this attractive wood. It offers wonderful views across Clarach Bay and a picnic spot to relax. From there it is possible to continue east to Coed y Cwm [109] and so continue with a circular walk back to town. Alternatively, continue north along the coast path to reach the unique wonder of the petrified trees at Borth Forest [106]. The nearest car park (Access point) is at Clarach Bay.

110 The Forest Guide: Wales

BORTH FOREST

BORTH, CEREDIGION   MAP REF 106

Ownership: Public: Crown Estate
Designations: SSSI
Area: 0ha
Forest type: petrified

Forest location: SN607911
Explorer Map: OL23
Ease of access: Easy

Access point: SN6062692561
///journey.tracking.rewrites
SY24 5JU

Situated between the raised peat bog of Cors Goch Fochno NNR and the sparkling waters of Bae Ceredigion/Cardigan Bay, the 6km-long stretch of Borth sands has no living trees. Yet, appearing and disappearing with tides and shifting sands, the stumps and roots of trees can sometimes be seen. What may look at first to be a rock will reveal distinct tree rings when the fronds of seaweed are carefully parted. This is a petrified forest, which thrived about 5,000 years ago. Dendrochronologists have identified them as the remains of birch, hazel, oak, Scots pine and willow. Human and animal footprints have also been discovered by archaeologists, along with signs of ancient hearths, tools and even a raised walkway constructed with wattle (woven coppiced timbers). Legend

suggests this is the mythical kingdom Cantre'r Gwaelod, which was flooded after Seithenyn, a drunken prince, failed to shut the town's gates. Some say the warning peals of the church bells of Cantre'r Gwaelod can be heard on a stormy night.

Visit at low tide. The northern end of the beach reveals more of the forest, especially around 1km south of the Access point.

Above: The petrified remains of Borth Forest [106].
Below: Clarach Bay.

GOGERDDAN WOOD

PENRHYN-COCH, CEREDIGION — MAP REF 108

Ownership: Public: Natural Resources Wales
Designations:
Area: 34ha
Forest type: mixed
Forest location: SN630841
Explorer Map: 213
Ease of access: Easy/Moderate
Access point: SN6344383760
///regress.extremes.sailing
SY23 3ED

Situated next to a major environmental research institute linked with Aberystwyth University, this medium-sized woodland offers a peaceful visiting experience. It is named after a nearby mansion, once owned by a family of lead and silver mine owners.

Many ancient oak, lime and sweet chestnut trees grow on the hillside, which in spring is carpeted by bluebells. A 2.4km circular waymarked trail starts at the car park (Access point), which has a small picnic area.

COED Y CWM

ABERYSTWYTH, CEREDIGION — MAP REF 109

Ownership: Public: Ceredigion County Council
Designations: LNR
Area: 30ha
Forest type: broadleaved
Forest location: SN599834
Explorer Map: 213
Ease of access: Easy/Moderate
Access point: SN6022583208
///trinkets.bloodshot.gazes
SY23 3DG

Trees have grown on the steep slopes above Clarach village for hundreds of years. The semi-natural ancient woodland is dominated by ash and sycamore, but also supports beech, hornbeam, sessile oak, small-leaved lime, sweet chestnut and whitebeam and some Scots pine trees. Among the bramble and honeysuckle lurks enchanter's nightshade, rare bird's-nest orchid and a rich layer of ferns. The LNR includes an old quarry and shady ancient mine workings.

From a small lay-by on the B4572 (Access point), climb the steps to enter the wood. Keen walkers could begin a stunning 6km-long circular walk from Aberystwyth, following the coast path north up Constitution Hill before passing through the neighbouring wood of Glanmor Fach [107] to reach Coed Y Cwm, and returning via Coed Penglais [111].

MYHERIN FOREST

PONTARFYNACH/DEVIL'S BRIDGE, CEREDIGION — MAP REF 110

Ownership: Public: Natural Resources Wales
Designations:
Area: 1,171ha
Forest type: conifer
Forest location: SN783773
Explorer Map: 213
Ease of access: Difficult
Access point: SN7713981417
///dentistry.kilt.mixer
SY23 3LE

The Mynyddoedd Cambrian/Cambrian Mountains are often described as a green desert, the remote uplands being little visited and offering a rare sense of the wild in Britain. Located east of Pontarfynach/Devil's Bridge, Myherin Forest is unnamed on many maps, yet this large, productive coniferous forest reveals many secrets for those who are well equipped and capable of navigating safely outdoors.

At its most northern tip, Rhaeadr Myherin is one of Wales' most remote and least-visited waterfalls. The Access point is a small lay-by next to the junction of the B4343 with the A44. Look for the footpath about 50m north-east, which leads down

Water tumbles through the trees at Myherin Forest [110].

to a footbridge. The start of the walk is rather unattractive, passing by an old lead mine and its tailings. The route soon climbs steadily, and it is a demanding walk south-east over steep and wet ground to reach the ridge offering dramatic views of the huge forest. Several streams tumble from the high ground, while the most dramatic falls lie another kilometre to the east (SN7972180712 | ///hitters.fracture.corkscrew). The trees grow at the upper altitudinal limit for conifers in Wales (almost 500m above sea level), and many are stunted or windblown due to the exposure.

It is possible to descend into the valley of the Afon Myherin and follow its channel westward, though this is a very challenging route. The valley is the experimental site of many novel 'leaky dams', created using brash (branches removed from trees) to help slow the flow during flood events.

Deep within the remote Myherin valley lies Nant Syddion (SN7734179061 | ///trace.collision.using), one of only nine bothies in Wales. The farmhouse is large by usual bothy standards, yet its location and history lend it an eerie quality. In 1856, the first quadruplets recorded in Wales all died within a week or so of being born here to the Hughes family. Tragically, within a month Margaret Hughes lost two other children and her husband to a mystery illness. The easiest access for the bothy (6km return) is to start at Tymawr Farm campsite (SN7552779282 | ///instead.topples.financial) which, when open, offers a small car park with an honesty box for payment. Alternatively, follow a 9km return route from the south, starting at Coed yr Arch [119]. Overall, the Coed yr Arch access point offers an easier option to explore part of Myherin Forest. The Pontrhydfendigaid Trail and Cambrian Way also both pass through the forest.

COED PENGLAIS

ABERYSTWYTH, CEREDIGION   MAP REF 111

Ownership: Public: Ceredigion County Council
Designations: LNR
Area: 11ha
Forest type: broadleaved

Forest location: SN591821
Explorer Map: 213
Ease of access: Easy/Moderate/Difficult

Access point: SN5876281848
///meal.salary.opts
SY23 2BF

Coed Penglais [111] seen from Aberystwyth.

Supported by an active local group of volunteers, this small LNR is located immediately above the town of Aberystwyth, offering stunning views across Bae Ceredigion/Cardigan Bay. Once part of the Penglais Estate, stands of old beech trees and stone walls remain from the eighteenth century. A quarry to the west of the site provided stone for local house-building, and now harbours ferns, invertebrates and reptiles. Park in town to reach the wood's entrance on the junction of Infirmary Road with North Road (Access point). The wood can be included as part of an attractive circular walk (see Coed y Cwm [109]).

BWLCH NANT YR ARIAN FOREST

GOGINAN, CEREDIGION   MAP REF 112

Ownership: Public: Natural Resources Wales
Designations: SM
Area: 756ha

Forest type: conifer
Forest location: SN715818
Explorer Map: 213
Ease of access: Moderate

Access point: SN7180181382
///screaming.urge.tulip
SY23 3PG

Offering fine views of the Mynyddoedd Cambrian/Cambrian Mountains and the distant shimmering water of Bae Ceredigion/Cardigan Bay, this large forest is a major draw for active and sport-loving visitors. Networks of trails cater for walkers, mountain bikers, horse riders, runners and orienteers. Some parts of the forest also host rally car races.

The site has a long tradition for feeding the magnificent red kite, where a daily afternoon session regularly attracts spectacular numbers of birds (often 150 individuals), which can be watched wheeling over the trees or seen close up, thanks to a large bird hide.

At the western end of the forest, overlooking Goginan and the Melindwr Valley, are the remains of the Roman-era hill fort Banc-y-Castell (SN6939981851 | ///professed.basic.haven). Some historians believe this is also the site of a medieval timber castle, recorded in 1216 as Nant yr Arian.

The site is one of ten Cambrian Mountains Dark Sky Discovery Sites (see also Coed yr Arch [119]), offering some of the best views of the night sky anywhere in Europe.

The car park near Llyn Pendam at the north end of the forest provides a quieter visitor experience (SN7100683924 | ///shred.causes.vipers).

COED RHEIDOL

PONTARFYNACH/DEVIL'S BRIDGE, CEREDIGION　　　MAP REF 113

Ownership: Public: Natural Resources Wales **Designations:** IPA, NNR, SAC, SSSI **Area:** 166ha	**Forest type:** temperate rainforest **Forest location:** SN745783 **Explorer Map:** 213 **Ease of access:** Moderate/Difficult	**Access point:** SN7529479062 ///digs.laughs.dugouts SY23 3JS

There are few more dramatic and unspoilt woodlands in Britain than the temperate rainforest growing on the precipitous slopes of the Rheidol Gorge. Stretching for 5km, the main designated areas are found on both banks above Pontarfynach/Devil's Bridge, and downstream on the south bank. The woodland is rich with bryophytes such as tree lungwort, western earwort and featherwort mosses, and ferns including lady fern and the rare forked spleenwort. Otters live in the water, pine marten in the trees, while nine bat species feed on the rich insect life found in the woods and glades.

Most visitors head for the famous Mynach Falls and the Devil's Punch Bowl. Access to two very dramatic walks is strictly controlled by a private landowner using turnstiles and a payment system. Both routes are demanding, including no fewer than 675 steps, and are unsuitable for dogs, pushchairs and for people with health conditions. Parking is available (SN7419477048 | ///lion.spirit.daydream) or the site can be reached via the Vale of Rheidol Railway. The narrow-gauge railway, which climbs 200m along its 19km scenic route from Aberystwyth, was originally built to carry timber and lead ore but now carries tourists during summer months. The railway timetable can limit the choice of walking routes. If this option appeals, check the websites of Pontarfynach/Devil's Bridge Falls and the Vale of Rheidol Railway to plan your visit carefully.

Alternatively, for a quieter experience away from the tourists, head 2km further upstream to explore the northern end of the woodland, which can be accessed on foot from a lay-by of the A4120 at Ysbyty-Cynfyn (Access point). Walk around the back of the chapel and follow a sunken ancient track, heading towards the sound of the roaring water. Enjoy wonderful views across the Rheidol valley before reaching the trees, at which point the footpath descends steeply and unevenly to the Afon Rheidol. The rather disappointingly ugly footbridge crossing over the dramatic cascades is known as Parson's Bridge, but it offers a great view of the swirling current below the arching stems of sessile oak and rowan trees.

The footpath to Parson's Bridge.

Bryophytes

Temperate rainforests (p.54) can most easily be identified by the prolific growth of mosses, lichens and ferns that carpet and hang from every rock and tree branch, lending these rainforests an otherworldly air and a sense of tingling secrecy. The often ancient and gnarly trees help to conserve moisture for these special plants, especially those known as bryophytes.

Definition

Bryophytes (from the Greek, meaning 'moss plants') are a division of non-flowering plants comprising mosses, liverworts and hornworts. Sometimes known as 'lower plants', they are classified by botanists as evolving between algae, the simplest of plants. While ferns are also a division of non-flowering plants, they have xylem and phloem, and hence are 'higher' or 'vascular' plants. Ferns and flowering plants have leaves, while all bryophytes lack real leaves, even though they may sometimes appear to have leaf-like structures known as a thallus. Bryophytes also lack stems and roots, and are typically small compared to higher plants.

Bryophyte groups

There are more than 1,000 species of bryophyte found in Britain. Mosses grow in many forms, from upright to creeping, and are found in all parts of the forest, from tree branches and stems, to the boulders and streams beneath their canopies. Liverworts can have thin leaf-like structures like those of mosses ('leafy liverworts') and can be quite intricately lobed, or may be formed of green plates ('thallose liverworts'). Hornworts are a small group of about 150 species that look like thallose liverworts, but their structures that produce spores resemble a horn, in contrast to the capsules of liverworts and mosses.

Life-cycle

Bryophytes have two alternating generations, called the sporophyte and the gametophyte. Each has a different physical form. The sporophyte produces spores, whereas the gametophyte, or the thallose or leafy stage, is when the sex organs (gametes) are produced.

Habitats

Generally, bryophytes are also epiphytes, meaning that they derive moisture and nutrients from the air, rain and the host surface (e.g. rocks or tree trunks) but not soil. They rely on photosynthesis to create energy, and respond rapidly to available moisture. The rough and acidic bark of oak is particularly good for bryophytes, whereas the smooth bark of beech and hornbeam is not. In Wales, clusters of temperate rainforest rich with bryophytic growth can be found in western coastal regions and along some of the deep river valleys running inland.

Identification

Identifying a moss, hornwort or lichen is often quite tricky. A good field guide and high-quality hand lens (×20 magnification) is required, and sometimes a microscope. Not knowing a species should not prevent you appreciating its beauty and complexity. Those interested in these fascinating lower plants can find out more from the British Bryological Society.

PANT DA

CAPEL BANGOR, CEREDIGION   MAP REF 114

Ownership: Private: Wildlife Trust of South and West Wales **Designations:** **Area:** 4ha	**Forest type:** broadleaved **Forest location:** SN670789 **Explorer Map:** 213 **Ease of access:** Moderate	**Access point:** SN6717878838 ///packages.tentacles.risks SY23 3NA

Once managed as a coppice, this woodland's old oak stools were cleared in the 1950s and replaced with larch. The site has subsequently undergone several changes, most recently following its purchase by the local wildlife trust. Most of the conifers have now gone, replaced with native broadleaves. Some original coppice stools can still be found among the regenerating mixed broadleaves. This peaceful site is a good location for spotting woodland birds. Paths are steep and uneven.

COED SIMDDE LWYD

PONTARFYNACH/DEVIL'S BRIDGE, CEREDIGION   MAP REF 115

Ownership: Private: Wildlife Trust of South and West Wales **Designations:** IPA, NNR, SAC, SSSI **Area:** 36ha	**Forest type:** temperate rainforest **Forest location:** SN717786 **Explorer Map:** 213 **Ease of access:** Moderate	**Access point:** SN7127178687 ///lecturers.fearfully.boomer SY23 3NF

Situated in the heart of the beautiful Rheidol valley, this upland woodland brims with precious wildlife. Most of the wood growing on the steep south-facing slope is sessile oak with occasional birch, but it becomes more diverse at its eastern end, with alder, ash, small-leaved lime, sycamore, wild cherry and wych elm. Ground flora includes common cow-wheat, common violet, primrose and wood anemone. The site is particularly rich in bryophytes, notably the liverworts straggling pouchwort and greater whipwort. Nine species of fern are present, including Wilson's filmy fern. Look for signs of badger, which thrive here, and brown hare along the wood's margins. A whole host of woodland birds can be seen, notably pied flycatcher, redstart and wood warbler.

A footpath leads into the wood from the entrance to Glyn-Rheidol Farm (Access point), where there is room to park a single car.

COED PENGLANOWEN

ABERYSTWYTH, CEREDIGION   MAP REF 116

Ownership: Private: Wildlife Trust of South and West Wales **Designations:** HPG **Area:** 5ha	**Forest type:** broadleaved **Forest location:** SN612785 **Explorer Map:** 213 **Ease of access:** Moderate	**Access point:** SN6117978627 ///tender.leans.drummers SY23 4LX

Previously a woodland managed for shooting as part of a traditional estate, the site was purchased in 1978 by the local wildlife trust. Some very tall grand fir and giant redwood can be found among the ash, beech, holly, sessile oak, sycamore and wych elm. An area of cherry laurel and rhododendron once planted for game cover and ornament continues to challenge site managers, as these invasive shrubs cast a dense shade, limiting native ground flora. The wood follows the valley bottom, and its damp shady atmosphere favours lichens and fungi such as *Enterographa crassa*, whose interlocking brown and spotted thalli can be seen on the trunks of older trees. The site is a good place to watch woodland birds, including pied flycatcher, nuthatch, tawny owl and treecreeper. A small lay-by is the shared Access point for the adjacent site of Old Warren Hill [117].

OLD WARREN HILL

ABERYSTWYTH, CEREDIGION · MAP REF 117

Ownership: Private: Wildlife Trust of South and West Wales	**Forest type:** mixed	**Access point:** SN6117978627
Designations: SM	**Forest location:** SN615787	///tender.leans.drummers
Area: 8ha	**Explorer Map:** 213	SY23 4LX
	Ease of access: Moderate	

This steep-sided hill is a perfect example of a 'dingle', a term commonly used to describe such sites, particularly in Wales. Its many broadleaved tree species are accompanied by Norway spruce and Scots pine. At the western edge of the nature reserve is a small stream, while at the top of the promontory, hidden among the trees, is the largest single-ditch hill fort in Wales, dating from the Iron Age. Badgers make their homes in its ramparts. Watch for buzzard and red kite soaring overhead, while large numbers of raven gather in the autumn. The many windblown trees provide ideal places to spot fungi.

A steep circular path begins in a small lay-by (Access point). In the valley below is the neighbouring reserve of Coed Penglanowen [116].

ALLT BOETH

PONTARFYNACH/DEVIL'S BRIDGE, CEREDIGION · MAP REF 118

Ownership: Private: Woodland Trust	**Forest type:** mixed	**Access point:** SN7343477721
Designations:	**Forest location:** SN736777	///draining.gripes.wiggling
Area: 22ha	**Explorer Map:** 213	SY23 3NB
	Ease of access: Moderate	

Lying just outside the multi-designated site of Coed Rheidol [113] and above the steeper sides of the Rheidol Gorge, this interesting woodland avoids the crowds of visitors at Pontarfynach/Devil's Bridge. While less spectacular, it has its own virtues. It possesses a sense of the wild, and exploring the wood offers spectacular views across the valley. For millennia its slopes were clad in sessile oak and birch, until it was planted with conifers during the twentieth century, changing its nature for many generations.

The owner is in the process of restoring broadleaved woodland, encouraging a diverse forest by adopting continuous cover forestry techniques. The site lies close to a release location chosen for the reintroduction of pine marten, and the mixture of spruce and pine with broadleaves will provide perfect habitat for this rare mammal. The Access point is found at the end of an unclassified road, which is unmetalled for the last 250m.

Sessile oak leaves.

COED YR ARCH

PONTARFYNACH/DEVIL'S BRIDGE, CEREDIGION   MAP REF 119

Ownership: Public: Natural Resources Wales **Designations:** **Area:** 45ha	**Forest type:** mixed **Forest location:** SN765755 **Explorer Map:** 213 **Ease of access:** Moderate	**Access point:** SN7655575577 ///swarm.violinist.dialect SY23 4AB

It is almost impossible to miss the large stone archway next to the B4574 that lends this woodland its name. It was built on the occasion of the golden jubilee of King George III and was once the gateway to the Hafod [120] estate. Three waymarked trails lead from the car park (Access point), exploring the former grounds of the estate. Some magnificent 200-year-old beech trees can be seen, while routes that climb higher into the forested hills of Myherin Forest [110] provide wonderful panoramic views, including the distant peak of Cadair Idris in Eryri/Snowdonia National Park. The site is one of ten Cambrian Mountains Dark Sky Discovery Sites (see also Bwlch Nant yr Arian Forest [112]), offering some of the best views of the night sky anywhere in Europe.

BLACK COVERT

LLANILAR, CEREDIGION   MAP REF 121

Ownership: Public: Natural Resources Wales **Designations:** **Area:** 20ha	**Forest type:** mixed **Forest location:** SN670724 **Explorer Map:** 213 **Ease of access:** Easy	**Access point:** SN6670972907 ///cocoons.beards.vowed SY23 4AT

A 'covert' is an area managed for raising game birds, revealing that this small woodland was once owned and managed as a shooting resource by the Trawsgoed Estate. Now open to the public, the small riverside woodland is a very tranquil place. A central stand of conifers is surrounded by a ring of broadleaves, mostly beech, which comes to life with bluebells in the spring. A short surfaced and waymarked trail follows the fast-flowing Afon Ystwyth, passing under some impressively tall conifer trees. On the opposite riverbank lies Trawsgoed Fort. Discovered as recently as 1959, it is believed to have been a major Roman military base located next to a strategically significant river crossing. Another historical feature from the period is found to the west at nearby Coed Allt-Fedw [122].

Black Covert and the Afon Ystwyth [121].

The waterfall of Rhaeadr Peiran at Hafod [120].

HAFOD

CWMYSTWYTH, CEREDIGION

MAP REF 120

Ownership: Private: Hafod Trust
Designations: HPG, SM, SSSI
Area: 239ha
Forest type: conifer
Forest location: SN768734
Explorer Map: 213
Ease of access: Moderate/Difficult
Access point: SN7680573673
///receiving.marmalade.exclusive
SY25 6DX

Hafod was among the earliest tourist destinations in Wales, made popular by its most famous owner Thomas Johnes. Beginning in the late eighteenth century, he laid out the grounds in the fashionable picturesque style, attracting young men from across Britain who were unable to complete the Grand Tour of Europe owing to the outbreak of the Napoleonic Wars. During the nineteenth and twentieth centuries the estate fell into disrepair, but since the 1990s it has been lovingly restored. The estate is now managed jointly by the National Trust and National Resources Wales under a long-term lease.

The scenery is spectacular, and nature abounds. Otter and brown trout swim among the dark currents, and dipper, kingfisher and grey wagtail feed on the surface. Crossbills and pine marten live in the woods.

Some 14km of paths and five waymarked trails have been reinstated. These allow viewing of the dramatic falls of Rhaeadr Peiran and the Gothic Arcade overlooking the Afon Ystwyth. The 'Gentleman's Walk' is perhaps the most satisfying but is quite strenuous. A range of shorter and more accessible routes can be followed, including the popular 'Lady's Walk'.

COED ALLT-FEDW

LLANILAR, CEREDIGION | MAP REF 122

Ownership: Public: Natural Resources Wales
Designations: SM
Area: 116ha

Forest type: conifer
Forest location: SN657729
Explorer Map: 213
Ease of access: Moderate

Access point: SN6652672856
///giggled.retaliate.bravery
SY23 4AT

With commanding views of the winding Afon Ystwyth, it is no wonder that our Iron Age ancestors built a hill fort on the top of this prominent hill. Now clad with conifers, it has two main routes for visitors to enjoy through peaceful woodland, which is partly managed as a nature reserve. A level track curves northwards around the foot of the hill from the Access point, passing through mixed woodland. Alternatively, another can be followed westward where it climbs steadily, passing a pond buzzing with dragonflies and damselflies. Soon after, a ride climbs steeply to reach a clearing on the top of the hill. It is also possible to use the car park at nearby Black Covert [121] and follow a trail to reach this site.

TY'N Y BEDW

TYNYGRAIG, CEREDIGION | MAP REF 125

Ownership: Public: Natural Resources Wales
Designations: SAC, SSSI
Area: 741ha
Forest type: conifer

Forest location: SN712710
Explorer Map: 213
Ease of access: Easy/Moderate/Difficult

Access point: SN6943971630
///melon.register.engulfing
SY23 4BJ

This moderately sized conifer forest offers a range of peaceful walks with scenic views across the rolling wooded landscape, graced by soaring red kite and buzzard. Two waymarked trails provide a choice of routes but both require some steep ascents. A stand of magnificent Douglas fir trees grows further up the slope. In the valley bottom, near the pretty Afon Ystwyth, the fringes of the wood reach Grogwynion Nature Reserve, which can be accessed via a short boardwalk. Named after a former lead mine, the river gravel and shingle are full of heavy metals, yet harbour many interesting plants and invertebrates. Sustrans Route 81 passes the Access point.

COED Y BONT

PONTRHYDFENDIGAID, CEREDIGION | MAP REF 126

Ownership: Public: Natural Resources Wales
Designations: SM
Area: 27ha

Forest type: mixed
Forest location: SN735656
Explorer Map: 180
Ease of access: Easy/Moderate

Access point: SN7378665974
///sailed.flexed.sharpness
SY25 6EP

Coed y Bont is managed by the volunteers of a local community woodland association and embraces two adjoining woods. On the flat valley bottom near the Afon Teifi, Coed Dolgoed is a mix of young broadleaved trees, including aspen, downy birch, hazel and oak. To the south on a hillside, Coed Cnwch consists of a mix of species, including some old oak trees among the exotic conifers, which reveal its ancient origins. Easy trails can be followed in the valley bottom, accompanied by interpretation signs and benches, while another heads to the top of the hill, which offers good views. Nearby lies the remains of an Iron Age hill fort at Gilfach y Dwn Fawr.

The wood is designated as a Dark Sky Discovery Site thanks to low levels of light pollution and panoramic views of the heavens. Tawny owl is one of more than 80 bird species found on the site.

The area was once within the grounds of Strata Florida, one of the largest abbeys in Britain, and the impressive ruins nearby are worth visiting. Just beyond is access to the northern tip of Wales' largest forest at Tywi [127].

PENDERI CLIFFS

LLANRHYSTUD, CEREDIGION

MAP REF 123

Ownership: Private: Wildlife Trust of South and West Wales	Forest type: broadleaved	Access point: SN5589672967
Designations: SSSI	Forest location: SN552735	///airfields.concerned.tequila
Area: 2ha	Explorer Map: 213	SY23 5DP
	Ease of access: Moderate	

Chough and raven soar above the cliffs, which run for 2km along this special nature reserve. It is also a good place to spot peregrine falcon. The cliffs are clad in a unique scrub of small-leaved lime, while a small area of sessile oak woodland has a wide range of minor tree species, including blackthorn, hawthorn, rowan and wych elm. Dog's mercury, wood anemone and wood sorrel thrive under the trees, and ferns grow in sheltered gullies. Linnet, stonechat and whitethroat shelter and breed in the trees.

At the reserve's southern end, there is a good view down to the rocky platforms that appear at low tide, used by moulting grey seals during March and April.

The reserve is crossed by the Ceredigion Coast Path, which can be joined at various points from the north or south. The shortest access route is via a public footpath leading from the A487 (Access point), which conveniently starts just 50m south of a lay-by. Beware steep cliffs!

Penderi Cliffs [123].

COED MAENARTHUR

PONT-RHYD-Y-GROES, CEREDIGION   MAP REF 124

Ownership: Public: Natural Resources Wales **Designations:** SAC, SM, SSSI **Area:** 65ha	**Forest type:** mixed **Forest location:** SN730723 **Explorer Map:** 213 **Ease of access:** Moderate/Difficult	**Access point:** SN7383272255 ///means.neckline.returns SY25 6DQ

On a knoll overlooking the steeply wooded Ystwyth valley, Castell Grogwynion is one of the largest hill forts in Wales (SN7211572505 | ///circus.demoted.storyline). The rectangular defensive fort, which dates from the Iron Age, measures 170m by 100m. It is reached by following a spectacular walk through Coed Maenarthur, passing through areas of sessile oak and mixed conifers. The scenery is wonderful at any time of year, especially the dappled sunshine in summer and autumn colours. The waterfalls and rapids of the Ystwyth gorge are dramatic after heavy rain. Enjoy views through the trees to the top of the hill fort.

This area has a rich mining history, especially lead and other heavy metals, and spoil heaps can be seen throughout the valley. The eastern tip of Grogwynion Nature Reserve, reached from Ty'n y Bedw [125], extends into Coed Maenarthur.

Walk to the Access point from the town centre or park carefully in the lane. Look for a small gate, opposite the Lisburne Metal Mine Waterwheel, from where a path leads down to a wooden miner's bridge over the gorge. A 5km-long waymarked trail, which takes in both hill fort and gorge, is often steep with some long ascents and descents, plus occasional precipitous drops near the path. The trail's white arrows are designed only to work one way (anticlockwise) and are not always very clear, so a good-quality map is also advised.

Walkers head across the miner's bridge into Coed Maenarthur [124].

ALLT CRUG GARN

PENNANT, CEREDIGION　　　　　　　　　　　　　　MAP REF 128

Ownership: Private: Wildlife Trust of South and West Wales
Designations:
Area: 1ha
Forest type: broadleaved
Forest location: SN518617
Explorer Map: 199
Ease of access: Moderate
Access point: SN5175261725
///reward.ferried.plan
SY23 5JR

Natural succession is dynamically active on this small nature reserve. Old heathland at the north of the site is gradually being invaded by birch, rowan and willow, with occasional beech and spruce. Look for narrow buckler fern in shady areas, and emperor moth among heather during summer months. The inedible false truffle grows among birch leaf litter in the autumn, accompanied by the dark club-shaped fruiting bodies of the parasitic fungus *Cordiceps ophioglossoides*. Park at the reserve entrance (Access point) opposite the entrance to Grug Garn Farm but beware the soft verge.

LLANERCHAERON

ABERAERON, CEREDIGION　　　　　　　　　　　　　MAP REF 129

Ownership: Private: National Trust
Designations: HPG, SSSI
Area: 10ha
Forest type: broadleaved
Forest location: SN480603
Explorer Map: 198
Ease of access: Easy/Moderate
Access point: SN4810860209
///introduce.detect.upwards
SA48 8DG

Among the attractive parkland and larger wooded estate, the 10 hectares of Coed Allt Lan-las is an SSSI thanks to its old oak trees and rich ground flora. On the opposite side of the road from the elegant Georgian villa of Llanerchaeron, follow a range of woodland and riverside (Afon Aeron) trails from the car park (Access point). This is a Dark Sky Discovery Site due to the unspoilt night skies.

CWM BERWYN FOREST

TREGARON, CEREDIGION　　　　　　　　　　　　　MAP REF 130

Ownership: Public: Natural Resources Wales
Designations:
Area: 2,377ha
Forest type: conifer
Forest location: SN737573
Explorer Map: 180
Ease of access: Easy/Moderate
Access point: SN7376957378
///howler.outcasts.shell
SY25 6PH

The forest roads in this large coniferous plantation provide many kilometres of informal access, yet it is little visited. It is possible to walk to the forest from nearby Tregaron or Llanddewi Brefi, but the Access point described provides a convenient start to exploring the area while offering a good view over Llyn Berwyn. Like nearby Tywi Forest [127], among the blanket Sitka spruce, pockets of lodgepole pine provide valuable food to the endangered red squirrel. The Mid Wales Red Squirrel Partnership has monitored their population in the forest for several years, installing trail cameras and feeding stations.

TYWI FOREST

PONTRHYDFENDIGAID, CEREDIGION　　　　　　　MAP REF 127

Ownership: Public: Natural Resources Wales
Designations:
Area: 6,112ha
Forest type: conifer
Forest location: SN810559
Explorer Map: 187
Ease of access: Moderate
Access point: SN7555964601
///competent.little.surcharge
SY25 6ES

Sprawling across the southern end of the Mynyddoedd Cambrian/Cambrian Mountains, Tywi Forest is the largest forested area in Wales and among the most remote. Planted by the Forestry Commission in the early twentieth century for timber production, it extends from Pontrhydfendigaid in the north, to Beulah, Llanwrtyd, and Rhandirmwyn in the south. Divided in half by the Afon Tywi/River Towy, its trees straddle the counties of Ceredigion and Powys. At first glance, many of its plantations may appear to be sterile blankets of conifers, yet were it not for this forest's vast scale, the red squirrel may not have survived in Mid Wales. Most of the forest comprises Sitka spruce, but pockets of lodgepole pine provide perfect habitat for this endangered animal. Experimental field

Llyn Brianne reservoir and Tywi Forest [127].

trials of lodgepole pine were planted here in 1970, proving that this exotic species of pine thrives on wet and exposed ground that is disliked by almost all other tree species. It establishes rapidly and produces seed from an early age, which conservationists now realise also provides essential food for the red squirrel, while the spruce provides arboreal corridors for safe movement through the forest.

This remains a productive working forest, but walkers and cyclists can informally explore the very extensive network of forest rides. Given its scale, unsurprisingly there are many places that it can be accessed. Travelling between Tregaron (past Cwm Berwyn Forest [130]) and Llanwrtyd Wells on the unmarked road that dissects the forest is an unforgettable experience, offering a unique perspective of the awesome scale of this forest. A good place to stop and enjoy views across the trees and Llyn Brianne is at Carreg Clochdy (SN8121250152 | ///custodian.stammer.bachelor). For walkers and cyclists, a waymarked trail can be followed at the northern tip of the forest. Starting at the Access point, near to the thirteenth-century Cistercian Abbey at Strata Florida, it passes close by a medieval holy well.

For the experienced walker, there is even the option to stay overnight in the forest at Moel Prysgau (SN8060361124 | ///dorms.owner.hound), one of Wales' nine bothies.

COED ALLT CEFN MAESLLAN

LLANARTH, CEREDIGION MAP REF 131

Ownership: Private: Woodland Trust Designations: Area: 7ha	Forest type: broadleaved Forest location: SN426583 Explorer Map: 198 Ease of access: Moderate	Access point: SN4236457699 ///baguette.speaker.heads SA47 0NQ

This thin ribbon of trees grows along the pretty Afon Llethi, sandwiched between a static caravan park and the village of Llanarth. Few people, other than local residents or holidaymakers, visit the site, but it has much to offer. A footpath tracks the river, weaving between the sessile oak trees, many of which were once cut as coppice. At its northern end the slopes become less steep, with ash and alder becoming frequent just before a boundary where woodland ownership transfers to another private owner. The footpath continues, although making a circular route would involve some lengthy stretches of road walking. To reach the wood, park in the village of Llanarth and navigate to the start of a footpath (Access point) near the church, where the footpath sign is mounted high on a telegraph pole.

ALLT PENCNWC

YSTRAD AERON, CEREDIGION MAP REF 132

Ownership: Private: Wildlife Trust of South and West Wales Designations: Area: 4ha	Forest type: broadleaved Forest location: SN518556 Explorer Map: 199 Ease of access: Moderate	Access point: SN5187855713 ///thickened.nipped.allergy SA48 7PQ

Large spreading oak trees make up the canopy of this slither of woodland running alongside the north side of the Afon Gwili. Great spotted woodpecker, nuthatch and treecreeper feed on the trees, while bluebell, lesser celandine and wood sorrel colour the woodland floor. There are no paths within the wood, which grows on a steep slope. Park on the corner of the B4342 (Access point) and walk down the lane to enter the wood.

Wood sorrel.

LONG WOOD

LLANBEDR PONT STEFFAN/LAMPETER, CEREDIGION — MAP REF 133

Ownership: Private: Longwood Community Woodland
Designations: SM
Area: 132ha
Forest type: mixed
Forest location: SN603512
Explorer Map: 199
Ease of access: Easy/Moderate
Access point: SN6170052069
///glance.deed.rebounded
SA48 8LA

Formerly part of the public forest estate, Long Wood was purchased on behalf of the local community in 2003. Now the largest community woodland in Wales, Long Wood is managed by a small team of people including volunteers, supported by income derived from timber and firewood sales. As a PAWS site (see pp.188–89), the wood is being slowly transformed from being mainly coniferous to a mixed woodland, with many thousands of native tree species planted in recent years. A 15km network of trails has been created for walkers, horse riders and cyclists to enjoy. The remains of an Iron Age hill fort, Castell Goetre, overlook the wood and can be visited by following one of the bridleways. Although the wood grows along a fairly steep north-facing slope, many of the trails contour through the trees. A large car park provides easy access at the east end of the wood, while the trees can also be reached by walking from nearby Llanbedr Pont Steffan/Lampeter.

PARC TEIFI FOREST GARDEN

ABERTEIFI/CARDIGAN, CEREDIGION — MAP REF 134

Ownership: Private: Naturewise Community Forest Garden
Designations:
Area: 3ha
Forest type: broadleaved
Forest location: SN189463
Explorer Map: OL35
Ease of access: Easy
Access point: SN1882946244
///devalued.choirs.attend
SA43 1EW

Parc Teifi is run by a local community group as a forest garden following the principles of permaculture. Fruit and nut trees grow together in an intimate mixture, their needs carefully met through companion planting. As a forest garden, it also supports the well-being of local people. The site is only open while volunteers are active, usually Tuesdays every week and the third Saturday of a month. New volunteers are welcome. Check website for details.

COED Y FOEL

LLANDYSUL, CEREDIGION — MAP REF 135

Ownership: Private: Woodland Trust
Designations:
Area: 65ha
Forest type: broadleaved
Forest location: SN426424
Explorer Map: 185
Ease of access: Moderate
Access point: SN4266742061
///acoustics.rotation.somebody
SA44 4PD

Growing on the south-east slopes below the Iron Age hill fort of Pen Coed-Foel (254m), this oak woodland is a popular destination with local people. All the oak trees were felled during the First World War but have since regenerated as coppice with hazel, holly and rowan understorey. The woodland lights up with bluebells in spring, and provides a fruitful harvest of bramble and bilberry in autumn. A choice of routes can be followed from a small car park (Access point). Paths are often muddy and in places are steep.

Often defined as part of a larger region termed Mid Wales under modern political boundaries, the eastern part is defined by the county of Powys and termed Central-East in this guide. This region has the greatest forested area (more than 72,000ha) but being also the largest region, its forest cover is proportionally a relatively modest 14 per cent. It is the only landlocked region in Wales, and its border with England is rich with historic features, like the largest prehistoric hill fort in Wales at Llanymynech Rocks [137]. About one quarter of its southern area falls under the Bannau Brycheiniog/Brecon Beacons National Park, although it is sadly lacking in trees, except at sites like Craig Cerrig Gleisiad [191], Craig y Cilau [193] and Cwm Claisfer [192],

Looking back down the valley on the approach to Craig Cerrig Gleisiad [191].

where steep terrain deters grazing livestock. Other hill regions include Y Mynyddoedd Duon/Black Mountains, the Breidden Hills and the foothills of the Mynyddoedd Cambrian/Cambrian Mountains. Bryophytes thrive at Carngafallt [167] and Cefn Cenarth [160], while one of the largest areas of temperate rainforest in Wales falls within Coed Mellte [199] and Coed Nedd [200]. At Gilfach Nature Reserve [161] visitors watching the river may be lucky to spot an otter. Deeply incised wooded valleys, or dingles, have protected sites such as Cilcenni Dingle [182], Garth Dingle [181] and Cwm Byddog [180] from exploitation. The largest forest site is Dyfnant Forest [139], which, like Esgair Ychion [158] and Hafren [156], provides ample opportunities for recreation.

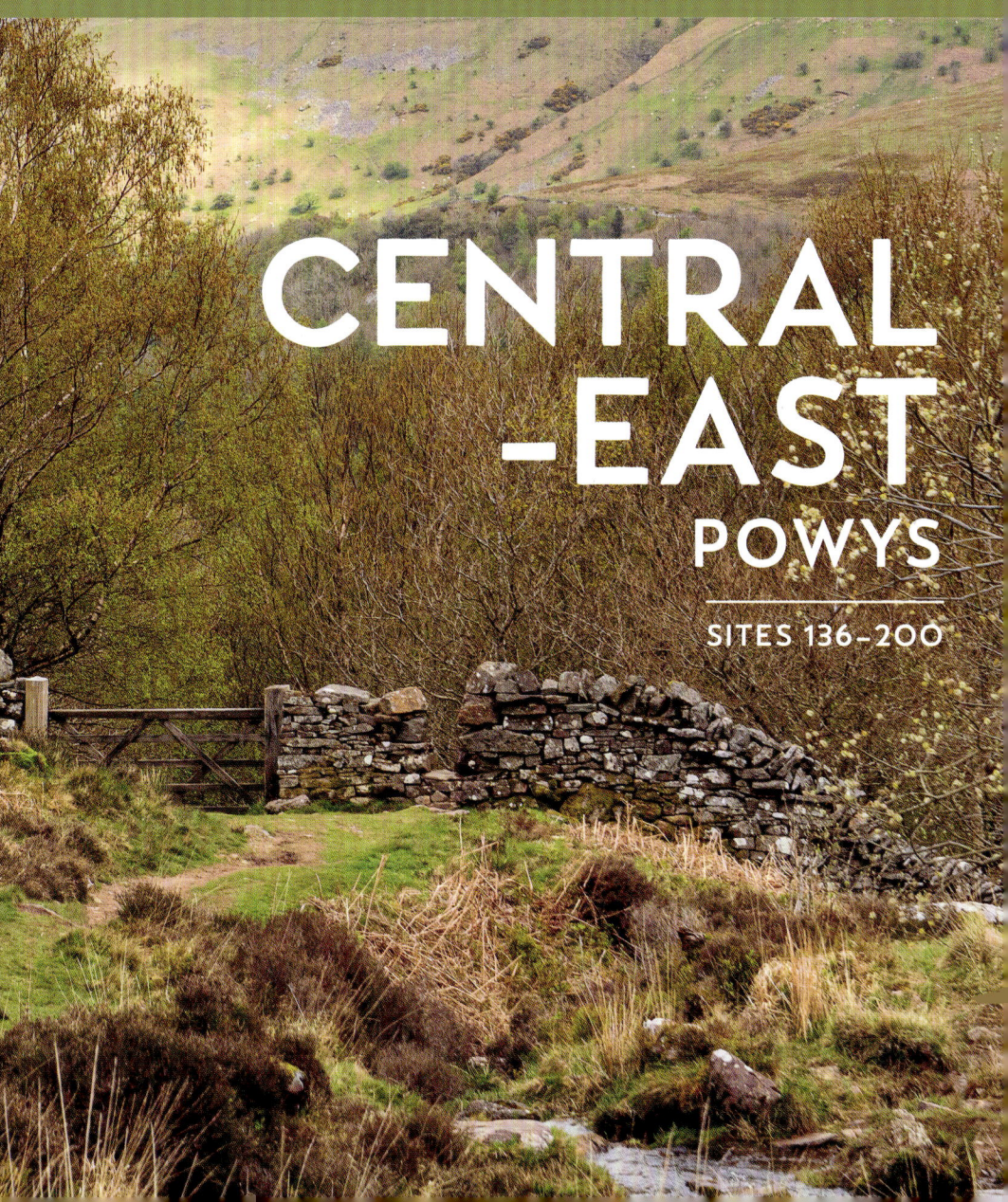

CENTRAL-EAST
POWYS
SITES 136–200

TAN-Y-PISTYLL

LLANRHAEADR-YM-MOCHNANT, POWYS MAP REF 136

Ownership: Private
Designations: SPA, SSSI
Area: 12ha
Forest type: broadleaved

Forest location: SJ073294
Explorer Map: 255
Ease of access: Moderate

Access point: SJ0762729367
///contained.landowner.fairly
SY10 0BZ

Birch, rowan and oak trees grow in this small woodland on the steep east-facing escarpment of the Berwyn Mountains. The eyes of most visitors, however, are drawn to Pistyll Rhaeadr, Britain's tallest single-drop waterfall (80m).

There is a car park (charged) next to a tea room, offering a fine view of the falls, while a lay-by 250m further down the hill (Access point) provides a free alternative. A short path reaches the base of the falls, while a footpath can be followed to the top of the waterfall, offering wonderful views down the valley.

Pistyll Rhaeadr and the trees of Tan-y-Pistyll [136].

COED PENDUGWM

PONTROBERT, POWYS MAP REF 141

Ownership: Private: Montgomeryshire Wildlife Trust
Designations: SSSI
Area: 3ha

Forest type: broadleaved
Forest location: SJ105142
Explorer Map: 239
Ease of access: Moderate

Access point: SJ1017714210
///bills.huts.exactly
SY22 5JF

Part of a larger woodland (12ha) designated as an SSSI, this smaller part is managed by the local wildlife trust. Its ancient sessile oak trees growing on either side of the Afon Nant-y-Pandy are home to dormouse, a rare species of fungus gnat and the dark-edged bee-fly, together with hosts of birds including grey wagtail, pied flycatcher, redstart and wood warbler. Bluebell and wood anemone brighten the woodland floor.

Park opposite a farm entrance (Access point) and follow a track into the wood. Routes are waymarked but are often rough and steep, and include two stream crossings. Climb further up the hill for beautiful glimpses across the Welsh countryside.

LLANYMYNECH ROCKS

LLANYMYNECH, POWYS

MAP REF 137

Ownership: Private: Montgomeryshire Wildlife Trust
Designations: SM, SSSI
Area: 6ha

Forest type: broadleaved
Forest location: SJ267217
Explorer Map: 240
Ease of access: Moderate

Access point: SJ2707821944
///umpires.niece.parrot
SY10 9QR

The high limestone plateau of Llanymynech hosts the largest prehistoric hill fort in Wales, covering some 70ha, while Offa's Dyke Path passes to the west. Straddling the Wales and England border, the hill has been mined for its minerals for millennia, its surface deeply scarred by quarry workings, spoil heaps and access ramps. Today, the impressive rock faces provide a popular recreational destination for climbers. Since being designated as a nature reserve in 1972, the quarries have regenerated into precious habitat. Ash dominates, along with sycamore and occasional larch trees. Wild clematis, or 'old man's beard', cloaks many of the trees. The surrounding grassland and glades throng with butterflies, including the pearl-bordered fritillary and grizzled skipper, and a rich plant life can be seen including bee orchid, yellow rockrose and wild thyme.

To the east, the reserve extends outside Wales, where it is managed by the Shropshire Wildlife Trust. Views across the countryside of Wales and England are spectacular, which can be enjoyed with the help of a viewpoint with a toposcope (SJ2644821591 | ///encroach.almost.boat). A small car park (Access point) can be found at the end of Underhill Lane, while the site can also be reached via public footpaths passing through Llanymynech Golf Club.

Looking across the valley below Llanymynech Rocks [137] during autumn floods.

LLYN EFYRNWY/ LAKE VYRNWY

LLANWDDYN, POWYS

MAP REF 138

Ownership: Private: RSPB **Designations:** SM **Area:** 1,086ha **Forest type:** mixed	**Forest location:** SJ015192 **Explorer Map:** 239 **Ease of access:** Easy/Moderate	**Access point:** SJ0158519256 ///outwit.escaping.disco SY10 0NA

Built by the Victorians to provide fresh water to Liverpool, Llyn Efyrnwy/Lake Vyrnwy is lined on both sides by forest plantations. Most are coniferous, but there are patches of broadleaves, including sessile oak and silver birch. Watch overhead for goshawk and peregrine, and among the trees for crossbill, siskin, pied flycatcher, redstart and tree pipit, and on the water for goosander.

A rare day-flying moth, the Welsh clearwing, thrives here. Despite its distinctive transparent wings and orange fanned tail, it remains very elusive. Look for characteristic holes on the sunny side of the trunks of old birch trees, created by the adults when they emerge from feeding as larvae on the trees' inner bark. Often their yellow-brown pupal cases remain stuck in the exit holes, best spotted in late summer.

Several well-marked trails provide an excellent choice of routes for all abilities. Along the north shore, where the Afon Cedig joins the reservoir, grows a stand of particularly impressive conifers, including towering noble fir, Douglas fir and giant redwood. Just to the west lies the deserted rural settlement of Bryn Gwyn.

At the top end of the lake, the Centenary Hide provides good views across the water, where otters can sometimes be seen. It can be reached via a 5km circular walk, or by using an alternative car park nearby (SH9635924092 | ///dugouts.declares.extremely).

Llyn Efyrnwy/Lake Vyrnwy [138].

DYFNANT FOREST

PONT LLOGEL, POWYS

MAP REF 139

Ownership: Public: Natural Resources Wales **Designations:** **Area:** 2,236ha	**Forest type:** mixed **Forest location:** SJ003150 **Explorer Map:** 239 **Ease of access:** Easy/Moderate	**Access point:** SJ0321115631 ///interacts.down.powers SY21 0QG

Lying west of the Mynyddoedd Cambrian/Cambrian Mountains and cladding the rolling hills south of Llyn Efyrnwy/Lake Vyrnwy, Dyfnant Forest forms part of the National Forest for Wales. It is a popular destination for horse riders and horse-drawn carriage drivers, their needs catered for with more than 100km of tracks ('Rainbow Trails') plus purpose-built facilities to help with handling horses and large vehicles. A car park at Pen y Ffordd (SJ0175413543 | ///himself.chaos.intruders) is the best location for equestrians, and for walkers who appreciate immediate access into the forest.

An alternative for walkers is to start in Pont Llogel (Access point) and follow the Pererindod Melangell Walk westward along the banks of the Afon Efyrnwy/River Vyrnwy into the forest. The ultimate destination of the long-distance path (24km for keen walkers) is Britain's oldest Romanesque shrine at the church of Pennant Melangell (SJ0242026553 | ///doctors.thickens.case). The ancient route, followed by pilgrims and drovers since the medieval period, is named after the patron saint of the hare and is waymarked by a symbol depicting a saint and the long-eared mammal.

The forest is host to a wide variety of woodland wildlife, but holds no conservation designations except for a small oak dingle that forms part of Dyfnant Meadows nature reserve and SSSI, managed by the Trefaldwyn/Montgomeryshire Wildlife Trust (SH999156). Anywhere in this large forest, keep an eye out for buzzard, great spotted woodpecker, cuckoo and goldcrest. With its mixture of broadleaves and conifers, it is a good place to hunt for fungi in the autumn, including waxcaps.

COED GLASLYN

FOUR CROSSES, POWYS

MAP REF 143

Ownership: Private: Woodland Trust **Designations:** **Area:** 11ha	**Forest type:** broadleaved **Forest location:** SJ042069 **Explorer Map:** 215 **Ease of access:** Difficult	**Access point:** SJ0344806877 ///interlude.scarecrow.soonest SY21 0HA

Coed Glaslyn is quite a remote high-altitude sessile oak wood, growing up to 300m above sea level. Formerly heavily grazed, as the surrounding land still is today, the trees were protected by the owners with a stock fence in 1986. Since then, the surviving large mature sessile oak, rowan and birch trees have been joined by young regenerating trees. Scattered large Scots pine trees are found occasionally, while some old pollarded beech trees grow at the northern end of the wood. The site drops away to the east, where wet woodland with downy birch and grey willow grows alongside rushes and sphagnum-rich mire. Keep an eye out for tree pipit displaying from the old oak trees and buzzard circling above. The Access point is reached after travelling along 4km of unclassified lanes, which end at a series of farm gates at the culmination of a spruce avenue. A public footpath heads east, passing close to the wood, which can be freely explored, though no paths are maintained. The site is often wet underfoot, and bracken growth is vigorous in summer months.

BREIDDEN FOREST

CREWGREEN, POWYS
MAP REF 140

Ownership: Public: Natural Resources Wales
Designations: IPA, SM, SSSI
Area: 197ha
Forest type: conifer
Forest location: SJ298145
Explorer Map: 240
Ease of access: Moderate
Access point: SJ2943414948
///twinkling.comedy.flow
SY5 9AZ

The Breidden Hills are not high in altitude compared to many others in Wales, but being surrounded by a gently rolling landscape, these three extinct volcanoes are very prominent landmarks lying between Y Trallwng/Welshpool and Shrewsbury. The forest-clad peak to the north of the group is Breidden Hill (365m). It boasts an impressive monument to Admiral George Rodney (1718–1792) who made his reputation in the American War of Independence, especially in 1782 at the Battle of the Saintes, where a combined French and Spanish fleet was defeated, preventing an invasion of Jamaica.

The main forest is coniferous but is being diversified to include more broadleaves. Keep a look out for crossbill in the canopy, nightjar in clearings and ride edges, and goshawk darting through the trees. Woodcock are often seen during winter months. Near the top of the hill the landscape is open with scattered oak trees. Look for tree pipit, kestrel and buzzard.

From the car park (Access point), follow tracks into the forest. At first, the route contours gently around the very steep northern flanks of the hill, before climbing up a narrow valley through the trees. It is a quite demanding ascent above the trees to reach the top of the hill, but the handsome reward is a spectacular view in all directions, including Llanymynech Rocks [137] on the opposite side of the broad valley of the Afon Hafren/River Severn.

The Admiral Rodney monument above Breidden Forest [140].

LLANBRYNMAIR FOREST

LLANGADFAN, POWYS　　　　　　　　　　　　　　　　MAP REF 142

Ownership: Private
Designations: SSSI
Area: 2,162ha
Forest type: conifer

Forest location: SH924099
Explorer Map: OL23
Ease of access: Difficult

Access point: SH9186313323
///bowls.passage.exhaling
SY21 0NY

One of the largest privately owned forests in Wales, Llanbrynmair stretches across a vast upland area, rising to the summit of Carnedd Wen (523m). The commercial forest, planted from 1960 onwards, is dominated (70 per cent) by Sitka spruce, and also contains three areas designated as SSSIs for their wet upland habitat. A large lake at its centre, Llyn Coch-hwyad, contains wild brown trout. Nearby, the dramatic yet remote waterfall of Llyn Coch-hwyad (SH9155710996 | ///king.apartment.rebel) receives very few visitors. Black grouse, honey-buzzard, long-eared owl, goshawk, nightjar and peregrine falcon might be seen anywhere in the forest by the sharp-eyed. Visitors should be experienced in navigating the outdoors as there are no waymarked trails, except Glyndwr's Way, which crosses an outlying block to the south of the forest (best accessed from Llanbrynmair).

COED DERI

LLANFAIR CAEREINION, POWYS　　　　　　　　　　　MAP REF 144

Ownership: Public: Llanfair Caereinion Town Council
Designations:
Area: 6ha

Forest type: broadleaved
Forest location: SJ099066
Explorer Map: 215
Ease of access: Easy/Moderate

Access point: SJ0989706527
///muddle.requiring.lush
SY21 0EE

Growing to the west of Llanfair Caereinion above the banks of the Afon Banwy, Coed Deri is an enjoyable wood to explore. A small footbridge crosses the wood at its western extremity, offering a chance to see brown trout and grayling beneath the surface, and goosander and kingfisher fishing from above. Nearer to town is Goat Field Arboretum, containing 25 native tree species and a modern stone circle. Listen for the distant whistle of the Welshpool and Llanfair Light Railway, now carrying tourists through the valley, but built originally to link local farming communities with Y Trallwng/Welshpool. The wood is readily accessible from town or from a small car park (Access point).

COED Y DINAS

Y TRALLWNG/WELSHPOOL, POWYS　　　　　　　　　MAP REF 145

Ownership: Private: Montgomeryshire Wildlife Trust
Designations:
Area: 4ha

Forest type: broadleaved
Forest location: SJ223052
Explorer Map: 216
Ease of access: Easy

Access point: SJ2207605174
///airbrush.disco.keepers
SY21 7AY

A ring of woodland and islet of trees surround the flooded old gravel pit, which supplied material for the bypass around nearby Y Trallwng/Welshpool. A hide on the edge of the trees provides the perfect place to watch for birds. Listen out for the *schip-schip-schip* of marsh tit in the branches of the alder and willow trees. The site's level paths are suitable for all abilities.

CHARLES ACKERS REDWOOD GROVE AND THE NAYLOR PINETUM

TRE'R LLAI/LEIGHTON, POWYS MAP REF 146

Ownership: Private: Royal Forestry Society
Designations: SM
Area: 10ha
Forest type: mixed
Forest location: SJ251041
Explorer Map: 216
Ease of access: Moderate
Access point: SJ2566505230
///golf.gangs.booms
SY21 8HU

The oldest living specimens of coast redwood in Britain thrive in this special woodland. Planted by John Naylor in 1857, 15 years after its first introduction from California, the huge trees tower at 40m tall or more, yet are considerably less than half the height of the tallest specimens of this record-breaking species, which can exceed 110m in their native range! One interesting specimen, which was windblown in a gale but remains very much alive, exhibits 'phoenix regeneration', whereby many of its branches have turned into trees emerging from the fallen bole of their 'parent'.

In 1931, new owner Charles Ackers planted more coast redwood trees, before donating the site to the current owners in 1958. The society created a pinetum on the site in 1961, planting more than 100 different conifer species.

Access to the car park (Access point) is restricted to members of the Royal Forestry Society, and unfortunately there are no

Coast redwood foliage.

suitable places to park nearby. However, the Offa's Dyke (SM) national path passes through this beautiful wood, providing a wonderful walk for those prepared to hike from Forden or Tre'r llai/Leighton.

COED DOLYRONNEN

ABERCEGIR, POWYS MAP REF 147

Ownership: Private: Woodland Trust
Designations:
Area: 3ha
Forest type: broadleaved
Forest location: SH831000
Explorer Map: 215
Ease of access: Moderate
Access point: SH8301900174
///maple.advantage.trumpet
SY20 8NY

Coed Dolyronnen is a place to savour the quiet peace of rural Wales, as few people ever visit this small sessile oak woodland. A small lay-by (Access point) with room for a single car lies next to the bridge over Nant Gwydol, whose fast waters flow to the Afon Dyfi/River Dovey. From there it is possible to explore some of the wood to the north-west, but the bulk of the trees grow on a very steep north-facing slope. The easiest way to appreciate the remainder is to walk up the quiet lane, passing many old ivy-clad oak trees and hazel bushes. The calls of great tit and chaffinch accompany the visitor before the view opens up towards the top of the lane.

GREGYNOG

TREGYNON, POWYS

MAP REF 148

Ownership: Private: Gregynog Trust (University of Wales) **Designations:** HPG, NNR, SSSI **Area:** 72ha	**Forest type:** broadleaved **Forest location:** SO082975 **Explorer Map:** 215 **Ease of access:** Easy/Moderate	**Access point:** SO0849897569 ///lemons.mistaking.mouse SY16 3PN

The Great Wood is just one of many silvan attractions on this beautiful estate, owned by the University of Wales and managed by a charitable trust. Gregynog Hall was home to the Davies sisters Gwendoline (1882–1951) and Margaret (1884–1963), renowned philanthropists and art collectors. Hundreds of ancient trees and dozens of veteran trees grow within the main wooded area west of the centuries-old hall, and in the wood pasture, where the girths of many oak trees exceed 5m. The beautiful gnarled trees with hollow branches and immense boles are home to lesser horseshoe bat, a great diversity of woodland birds (notably pied flycatcher, redstart and wood warbler), rare lichens including tree lungwort, hosts of invertebrates and fungi. An arboretum includes some magnificent specimens of cedar, copper beech, giant redwood and monkey puzzle. Several waymarked trails can be followed around the estate, with interpretation signs featuring notable wildlife, which may be spotted throughout the year. Parking (charge) is available on site (Access point), where there are also facilities for visitors.

Gregynog Hall and ancient oak trees.

FFRIDD WOOD

TREFALDWYN/MONTGOMERY, POWYS   MAP REF 149

Ownership: Private **Designations:** SM **Area:** 16ha **Forest type:** broadleaved	**Forest location:** SO219972 **Explorer Map:** 216 **Ease of access:** Easy/Moderate/Difficult	**Access point:** SO2205496613 ///elephant.happening.mash SY15 6HN

The medieval town of Trefaldwyn/Montgomery was founded in 1223, constructed around the base of cliffs that formed the foundations for a once impregnable castle whose construction was begun by King Henry III. On the slopes above is an even older hill fort dating from the Iron Age. Ffridd Faldwyn Camp was constructed during the Roman conquest of Wales. Today, broadleaves clad its steep eastern flanks, including many ancient beech trees, accompanied by ash, sycamore and oak. Climb to the top of the hill, passing through the double ramparts of the hill fort, for spectacular views eastwards over the Vale of Trefaldwyn/Montgomery (England) and beyond to the distinct plateau summit of Corndon Hill (513m), which lies in Wales.

The Wales–England boundary here is very convoluted.

Park in Trefaldwyn/Montgomery Castle car park (charge) and walk west up the lane for 100m to reach a field gate (Access point), where a footpath leads through a field and woods to the B4385. Walk north-west along the road for 100m to find the next footpath, which climbs steeply up towards the trees and the hill fort. Two different circular routes can be followed using public footpaths, the shortest of which returns via a lane to the start. A longer route can be followed to take in Town Hill and the war memorial. The scenery everywhere is truly wonderful. Keep a look out for soaring buzzard and red kite.

DOLFORWYN WOODS

ABER-MIWL/ABERMULE, POWYS   MAP REF 150

Ownership: Private: Montgomeryshire Wildlife Trust **Designations:** **Area:** 28ha	**Forest type:** mixed **Forest location:** SO163959 **Explorer Map:** 215 **Ease of access:** Easy/Moderate	**Access point:** SO1585995637 ///vampire.heaven.thrilled SY15 6JG

Uncommon plants such as herb-Paris are indicators of the ancient woodland that once grew across the whole hillside overlooking Aber-miwl/Abermule. Although a significant area was felled to make way for commercial forestry, a wide variety of tree species was planted, and some original wooded areas preserved. The wood is therefore an interesting mixture of tree species providing attractive colour throughout the year. The elusive dormouse thrives in the wood, but visitors are more likely to see or hear woodland birds such as the jay, nuthatch, pied flycatcher, redstart and wood warbler. The wood is a great place to search for fungi in the autumn. In summer, pearl-bordered fritillary and speckled wood butterflies grace the rides and glades.

Penny bun or cep.

ROUNDTON HILL

CHURCHSTOKE, POWYS   MAP REF 151

Ownership: Private: Montgomeryshire Wildlife Trust **Designations:** IPA, NNR, SM, SSSI **Area:** 2ha	**Forest type:** broadleaved **Forest location:** SO291948 **Explorer Map:** 216 **Ease of access:** Moderate	**Access point:** SO2925594630 ///depravity.grows.format SY15 6EN

Roundton Hill (370m) stands proudly overlooking the Vale of Trefaldwyn/Montgomery. No wonder that in prehistoric times, a defensive hill fort existed on its craggy summit. Later, baryte and lead was mined on its slopes, and today the adits provide important hibernation sites for bats, especially the lesser horseshoe. A ring of broadleaved trees and scrub can be explored on the hill's lower slopes, providing shelter and breeding grounds for redstart, pied flycatcher, whitethroat and yellowhammer. Together with the surrounding grassland and crags, the site supports more than 300 locally or nationally rare species. Climb to the top of the hill, which is often steep, to enjoy truly spectacular views across the Welsh and English countryside.

HENGWM FOREST

FORGE, POWYS   MAP REF 152

Ownership: Public: Natural Resources Wales **Designations:** **Area:** 434ha	**Forest type:** conifer **Forest location:** SN773933 **Explorer Map:** 215 **Ease of access:** Difficult	**Access point:** SN7786495097 ///option.fresh.hurtles SY20 8RR

This is a lonely forest in a remote valley; reaching the Access point requires a 6km drive south out of Machynlleth along a single-track road. From there, forest tracks can be followed on foot or cycle further up the valley. Near its head, the Afon Hengwm tumbles dramatically from the hills surrounding Pumlumon Fawr (752m). It is a rarely glimpsed waterfall, and gaining a good view requires a battle across steep uneven terrain. This site should be tackled by those experienced in exploring the outdoors. Enjoy the tranquillity, accompanied only by the mew of buzzard overhead and *tsp-tsp* of goldcrest in the Sitka spruce.

COED GWERNAFON

LLAWR Y GLYN, POWYS   MAP REF 153

Ownership: Private: Woodland Trust **Designations:** SAC, SSSI **Area:** 31ha	**Forest type:** temperate rainforest **Forest location:** SN924907 **Explorer Map:** 214 **Ease of access:** Moderate	**Access point:** SN9230390302 ///overhaul.wallet.lurching SY18 6NT

Ancient sessile oak trees nestle in an arc along this quiet valley in the eastern foothills of the Mynyddoedd Cambrian/Cambrian Mountains. Hard fern, beech fern and oak fern thrive in the sheltered atmosphere of the wood, which includes holly in its understorey. The acid soils are carpeted in bilberry, while lichens abound on rocks and trees, including straggling pouchwort.

Park in a small lay-by near the bridge over the Nant Cwmcidyn (Access point), and follow a level path northwards, tracking the river.

FRONDERW WOOD

CERI/KERRY, POWYS   MAP REF 154

| Ownership: Public: Natural Resources Wales
Designations: SM
Area: 35ha | Forest type: conifer
Forest location: SO172889
Explorer Map: 214
Ease of access: Moderate | Access point: SO1709988746
///warbler.poetry.airbase
SY16 4PG |

Just north of the wooded valley of Drefor Dingle, the trees of Fronderw Wood cover the western flanks of a large unnamed hill. An interesting circular route can be followed through the wood and around the hill, taking in several sites of archaeological interest and returning via the dingle. Within the wood (SO1715388924 | ///paid.evidence.artichoke), now cleared of conifers, are the remains of a late prehistoric enclosure marked by a single bank and prominent ditch.

HAFREN FOREST

LLANIDLOES, POWYS   MAP REF 156

| Ownership: Public: Natural Resources Wales
Designations:
Area: 2,875ha | Forest type: conifer
Forest location: SN857869
Explorer Map: 214
Ease of access: Easy/Moderate | Access point: SN8573786938
///mammal.named.abolish
SY18 6SY |

Named after the Welsh name for the River Severn, Hafren Forest blankets a large area of hills west of Llyn Clywedog, forming part of the National Forest for Wales. A range of waymarked trails are offered from the main car park (Access point), from an easy route along the cascades of the Afon Hafren, to a strenuous route that extends north-west to the source of the river on the high ground of Pumlumon.

An alternative is to park in a lay-by (SN8709189707 | ///acrobat.thickens.employ) offering scenic views across Llyn Clywedog. From there, a circular route can be followed around a promontory. The main advantage of this option is its proximity to the reservoir and the greater chance of spotting an osprey during summer months. Thanks to dedicated efforts by conservationists, this beautiful bird of prey has been saved from the brink of extinction.

Hafren Forest [156] can be seen from far and wide.

PEN-Y-CASTELL WOOD

TREFEGLWYS, POWYS

MAP REF 155

Ownership: Private
Designations: SM
Area: 41ha
Forest type: broadleaved

Forest location: SN979882
Explorer Map: 214
Ease of access: Easy/Moderate/Difficult

Access point: SN9838787746
///wildfires.vaccines.woes
SY18 6LS

Pen-y-castell Wood wraps around the north side of Llyn Ebyr, a 12ha natural lake. The wood gains its name from a medieval motte and bailey castle, which sits atop the hill to the north-east. The views through the trees, across the lake to the distant Bryn y Fan (482m) are very attractive. A largely untrodden circular walk can be followed through the oak-dominated woodland and around the lake using public footpaths, passing through nearby Smith's Coppice and returning through the trees along the south shore of the lake. The lake is privately owned and used for coarse fishers to catch perch, pike and roach.

CERI FOREST

CERI/KERRY, POWYS

MAP REF 157

Ownership: Public: Natural Resources Wales
Designations:
Area: 423ha

Forest type: conifer
Forest location: SO153866
Explorer Map: 214
Ease of access: Easy

Access point: SO1487586298
///frame.relegate.offshore
SY16 4PE

This coniferous forest consists of several discrete blocks, stretching along the Kerry Ridgeway, offering fine views over the countryside of Wales, and England, which lies just 2.5km to the east as the crow flies. The area is rich in history. The car park in Block Wood (Access point) provides access to the Kerry Ridgeway path (24km), which is accessible to walkers, cyclists and horse riders. The ancient trackway was used by market traders and drovers to cross the border, linking the village of Dolfor with Bishops Castle in Shropshire, England. A prehistoric round barrow lies within the wood, while to the west beyond Ceri/Kerry Hill are two Bronze Age burial mounds known as Two Tumps.

ESGAIR YCHION

LLANGURIG, POWYS   MAP REF 158

Ownership: Public: Natural Resources Wales **Designations:** SM **Area:** 716ha	**Forest type:** conifer **Forest location:** SN854800 **Explorer Map:** 214 **Ease of access:** Moderate/Difficult	**Access point:** SN8430677282 ///script.anchovies.discrepancy SY18 6RS

Lying south of the upper reaches of the Afon Gwy/River Wye, this large conifer plantation is used by mountain bikers, and occasionally for car rallying. The area west of the Afon Diliw falls into Ceredigion, including a windfarm and the remote bothy of Nant Rhys (SN8367879271 | ///recline.opened.again). The Access point is reached at the end of a 5km-long unclassified road. Few people are encountered here.

In a clearing at the north-west of the forest is the Roman fort of Cae Gaer (SN8236381902 | ///volcano.difficult.anyway).

Historians believe it was possibly a temporary camp constructed to support the Roman army as it pushed westwards. Given the distance, use an alternative start point by parking in a lay-by (SN8227382453 | ///takers.noticing.pocketed) on the A44. Follow a footpath south to the Afon Tarennig, which must be crossed without the aid of a bridge. Experienced and fit walkers might consider continuing to reach Myherin Forest [110] and the dramatic falls of Rhaeadr Myherin.

TYLCAU HILL

MOELFRE CITY, POWYS   MAP REF 159

Ownership: Private: Radnorshire Wildlife Trust **Designations:** **Area:** 8ha	**Forest type:** broadleaved **Forest location:** SO138765 **Explorer Map:** 214 **Ease of access:** Moderate	**Access point:** SO1340076500 ///absent.purple.headers LD1 6UN

The traditional Welsh farmland with flower-rich pastures, hedgerows and dingle woodland surrounding Moelfre City (actually a hamlet!) are now managed as a nature reserve. Look for linnet and yellowhammer in the hedgerow trees of hawthorn, with occasional crab apple and rowan. Redstart and pied flycatcher breed in the woodland, and in the spring, calls of cuckoo echo through this tranquil valley. In the brooks, brown trout and bullhead fish attract otters. A waymarked 2km-long circular trail can be followed around the reserve, which can be steep in places. The route can be extended to reach Tylcau Hill (486m) to enjoy stunning views.

CEFN CENARTH

PANT-Y-DWR, POWYS   MAP REF 160

Ownership: Private: Radnorshire Wildlife Trust **Designations:** **Area:** 17ha	**Forest type:** broadleaved **Forest location:** SN966759 **Explorer Map:** 214 **Ease of access:** Moderate	**Access point:** SN9645675975 ///emblem.payout.blush LD6 5LL

Its name meaning 'ridge of lichens', this sessile oak woodland with aspen and downy birch offers a peaceful, scenic experience among wildlife, notably more than 100 lichen and 40 moss species. The yellow-green colonies of the bearded lichen *Usnea florida* are easy to spot, especially its 'lobes', which radiate from central fruit discs like clusters of flowers. The oak here was once coppiced for use in a local tannery as its bark is rich with tannins. In the twentieth century, conifers were planted on the upper slopes, but these have now been felled. Follow a waymarked trail, which zigzags quite steeply up through the trees. At the top, enjoy scenic views across the valley to the Bryn Titli windfarm and the Mynyddoedd Cambrian/Cambrian Mountains.

RADNOR FOREST

BLEDDFA, POWYS — MAP REF 162

Ownership: Public: Natural Resources Wales
Designations:
Area: 1,280ha
Forest type: conifer
Forest location: SO194660
Explorer Map: 200 and 201
Ease of access: Moderate
Access point: SO1887068279
///piston.radiated.closes
LD7 1PA

Once a royal hunting forest, and more recently designated as part of the National Forest for Wales, Radnor Forest consists of several wooded areas between Penybont and Llanandras/Presteigne. The largest of these lies south of Fishpools, where there is a generous car park (Access point). A range of informal routes can be followed, or a 4km-long waymarked trail, which includes two viewpoints. Marked on OS maps as 'Observatory' (SO19044467944 | ///duplicity.cries.quaking) is a stone tower constructed by the Birmingham Water Corporation to help with the sighting of the pipeline between the reservoirs of the Elan Valley (e.g. Penbont Woods [164]) and the English city they were built to supply.

Other sites within Radnor Forest include Burfa Bank [171], Nash Wood [169] and Warren Wood [175].

RHAYADER TUNNEL AND EMBANKMENT

RHAEADR GWY/RHAYADER, POWYS — MAP REF 163

Ownership: Private: Radnorshire Wildlife Trust
Designations:
Area: 2ha
Forest type: broadleaved
Forest location: SN963673
Explorer Map: 200
Ease of access: Easy
Access point: SN9663467784
///indicated.kind.outbound
LD6 5AS

Stretching along the old track of the Elan Valley Branch line of the former Mid-Wales Railway, this long, narrow nature reserve offers a fine walk with great views of the countryside as it heads up the Elan Valley. Just 500m south-west of Rhaeadr Gwy/Rhayader town the track enters an old tunnel, now an important bat hibernaculum, where five species roost and hibernate: Daubenton's, Natterer's, lesser horseshoe, brown long-eared and whiskered. A small gap left at the top of the brickwork, which blocks up most of the tunnel entrance, provides space for the many thousands of bats to access the tunnel. They feed on the insects that thrive on the flower-rich embankments and scrubby woodland.

From town, follow the Elan Valley Cycle Trail (also Sustrans Route 8 and 81) marked by a wooden gated entrance (Access point).

A Daubenton's bat flying at dusk.

The waters of Afon Marteg swollen by late summer rain at Gilfach Nature Reserve [161].

GILFACH NATURE RESERVE

RHAEADR GWY/RHAYADER, POWYS   MAP REF 161

Ownership: Private: Radnorshire Wildlife Trust **Designations:** SAC, SPA, SSSI **Area:** 14ha	**Forest type:** broadleaved **Forest location:** SN968718 **Explorer Map:** 200 **Ease of access:** Easy/Moderate	**Access point:** SN9647471686 ///oven.passively.ridge LD6 5LF

Upon reaching the beautiful Marteg Valley, visitors may consider themselves intrepid travellers, but they might think again. The leaping salmon, which can be watched from a viewing platform, have migrated here from the Atlantic Ocean, travelling through the Bristol Channel to swim up the Afon Gwy/River Wye, passing through England before reaching Wales. The upper reaches of the Afon Marteg are their final destination, where they will spawn at last.

The best time to see this spectacle is early to mid-November, when the Afon Marteg is swollen with autumn rains. Patient watchers may also be lucky enough to spot an otter. A car park near the confluence of the Afon Gwy/River Wye and Afon Marteg (SN9530871503 | ///ounce.voter.captions) is the easiest option to reach the waterfall, following the Wye Valley Walk eastwards. Grey wagtail, dipper and willow tit are frequently seen near the tree-lined water.

Further upstream, the wildlife trust has built a discovery centre in the traditional farm buildings where there is more parking (Access point) and a range of trails to follow, including the Wye Valley Walk. The woodland above the discovery centre is mostly sessile oak, brimming with birds, butterflies, lichens (look out for the unusual witches whiskers), mosses, fungi, bats and a rich ground flora. It is reached via the 'Oak Wood Walk'; a circular trail that follows an ancient trackway. The scenery is truly spectacular, and a match for the rich wildlife found across this wonderful place.

PENBONT WOODS

RHAEADR GWY/RHAYADER, POWYS — MAP REF 164

Ownership: Private: Elan Valley Trust
Designations: IPA, SAC, SPA, SSSI
Area: 115ha
Forest type: mixed
Forest location: SN913674
Explorer Map: 200
Ease of access: Easy/Moderate
Access point: SN9149167309
///lighters.push.coasters
LD6 5HE

The beautiful Elan Valley has five reservoirs, first constructed by the Victorians to provide drinking water for the English Midlands. The three central reservoirs include wooded areas, Penbont Woods near Penygarreg Reservoir being the highest in the valley. Among the conifer plantations are some magnificent tall specimens of Sitka spruce and Douglas fir, reaching almost 60m tall. The mostly broadleaved woodland beneath the dam, dominated by sessile oak, can be explored via two waymarked trails, which include some sections of steps.

WITHY BEDS

LLANANDRAS/PRESTEIGNE, POWYS — MAP REF 165

Ownership: Private: Radnorshire Wildlife Trust
Designations: SSSI
Area: 1ha
Forest type: broadleaved
Forest location: SO310649
Explorer Map: 201
Ease of access: Easy
Access point: SO3093465066
///lime.mystified.maple
LD8 2AX

A 'withy bed' is an Old English term for an area of coppiced willow, supplying rods for basket making, thatching, kindling and other uses. This small area of wet woodland straddles the Wales–England border, with the town lying immediately to the south and the beautiful River Lugg to the north. The tiny island of willow, alder and ash throngs with birds. Listen for the high calls of willow tit and the darting flights of feeding spotted flycatcher. Large red damselflies hunt over the water, where brown trout wait for mayflies.

The reserve is easily reached on foot from Llanandras/Presteigne, or use the car park next to the B4355 (Access point) and walk through the meadow, following an old leat to reach the reserve. A circular boardwalk caters for all abilities.

CORS ABERCAMLO

Y GROES/CROSSGATES, POWYS — MAP REF 166

Ownership: Private: Radnorshire Wildlife Trust
Designations:
Area: 3ha
Forest type: broadleaved
Forest location: SO073649
Explorer Map: 200
Ease of access: Easy
Access point: SO0729164995
///clogging.arranged.distilled
LD1 6RG

The birch and willow trees at Abercamlo provide shelter and habitat to wildlife, but are carefully controlled by reserve managers to prevent natural succession, which would result in the loss of precious bog habitat. Several species of sphagnum moss grow here, plus the carnivorous sundew. In the spring, look for frogspawn on the surface of the moss. Blackcap and garden warbler sing from the trees in early summer. The site is wet all year round, and has only one section of boardwalk, so suitable footwear is essential.

CARNGAFALLT

RHAEADR GWY/RHAYADER, POWYS   MAP REF 167

Ownership: Private: RSPB
Designations: IPA, SSSI
Area: 6ha
Forest type: broadleaved

Forest location: SN936651
Explorer Map: 200
Ease of access: Moderate

Access point: SN92874 64665
///absorbing.donates.endlessly
LD6 5HW

Immediately adjacent to Cnwch Wood [168], and also accessed from the Elan Valley Visitor Centre (Access point, charge), this small reserve is rather low-key, with no obvious site entrance. Only when inside the wood will the visitor come across an interpretation panel. The site is bounded by the valley road and the unclassified road that climbs Cwm yr Esgob, where towards the top of the hill (SN93405 64633 | ///folds.regrowth.gained) stands one of the red waymarkers that indicates a circular route through the woodland. Few people explore the reserve, where a great diversity of woodland birds can be seen, including pied flycatcher, nuthatch, redstart, lesser spotted woodpecker and whinchat.

CNWCH WOOD

RHAEADR GWY/RHAYADER, POWYS   MAP REF 168

Ownership: Private: Elan Valley Trust
Designations: IPA, SAC, SSSI
Area: 14ha

Forest type: broadleaved
Forest location: SN931648
Explorer Map: 200
Ease of access: Easy/Moderate

Access point: SN92874 64665
///absorbing.donates.endlessly
LD6 5HW

Below the dam of Caban-coch reservoir, the sessile oak trees of Cnwch Wood are easily accessible from the Elan Valley Visitor Centre (Access point, charge) via a level trail. For the more adventurous, climb up through the trees to reach the top of the dam, before walking south along the shores of the reservoir. Keep an eye out for red kite circling above and waterfowl on the open water. Another unnamed oak woodland grows under Craig Cnwch. Further still is the conifer plantation of Gro Wood [172].

NASH WOOD

LLANANDRAS/PRESTEIGNE, POWYS   MAP REF 169

Ownership: Public: Natural Resources Wales
Designations:
Area: 23ha
Forest type: mixed

Forest location: SO311633
Explorer Map: 201
Ease of access: Easy/Moderate/Difficult

Access point: SO31656 63513
///luggage.scribble.gathers
LD8 2LA

Several named woods including Nash Wood make up a sizeable area of mixed woodland south of Llanandras/Presteigne, forming part of the larger area of Radnor Forest. Nash Wood straddles the border between Wales and England. A 3.5km-long circular trail can be followed, which climbs quite steeply from the car park near the Access point. Look for crossbill, goldcrest and siskin in the tall Douglas fir and noble fir trees. A viewpoint takes in the beautiful scenery of the Radnor valley, including Burfa Bank [171], and rural Herefordshire in England.

LLANERCHI WOOD

RHAEADR GWY/RHAYADER, POWYS   MAP REF 170

Ownership: Private: Elan Valley Trust **Designations:** IPA, SAC, SSSI **Area:** 91ha	**Forest type:** mixed **Forest location:** SN908635 **Explorer Map:** 200 **Ease of access:** Moderate	**Access point:** SN9092763894 ///reverses.views.spits LD6 5HE

Llanerchi Wood lies at the south-west of Garreg Ddu Reservoir. Its steep sides have proved ideal for a range of adrenalin-pumping mountain bike routes. A series of pretty waterfalls tumble through the forest at Cwm Coel, where broadleaves dominate. Many of the conifer areas are being naturally regenerated with mixed stands. Look for long-tailed tit, song thrush and nuthatch among the oak, birch and rowan trees, and crossbill and siskin in the conifers. A 10km-long circular route can be followed around the reservoir, reaching Penbont Woods [164] before returning along the east shore and crossing the road bridge back to the Llanerchi car park (Access point).

BURFA BANK

LLANANDRAS/PRESTEIGNE, POWYS   MAP REF 171

Ownership: Public: Natural Resources Wales **Designations:** SM **Area:** 33ha	**Forest type:** conifer **Forest location:** SO282610 **Explorer Map:** 201 **Ease of access:** Moderate/Difficult	**Access point:** SO2778960960 ///shameless.shops.prompt LD8 2SA

Hidden among the trees on a prominent domed hill, Burfa Bank is the site of a large Iron Age hill fort, half-encircled by the Wales–England border. Dating from the Roman conquest of Wales (800 BC–AD 74), it measures 600×180m, with at least one defensive bank around its perimeter. There are no waymarked routes to help visitors in reaching the top of the hill. Look for a narrow path just beyond the Access point, and be prepared for a steep climb and some bramble bashing to find a clear way through during summer months. Glimpses through the trees of the expansive landscape beyond explain why this was once such an important place. For a great view of the hill itself, visit Nash Wood [169].

Bramble.

PENTROSFA MIRE

LLANDRINDOD/LLANDRINDOD WELLS, POWYS   MAP REF 174

Ownership: Private: Radnorshire Wildlife Trust
Designations:
Area: 2ha

Forest type: broadleaved
Forest location: SO060597
Explorer Map: 200
Ease of access: Moderate

Access point: SO0657760187
///asterisk.undulation.drones
LD1 5NY

The stream in this quiet valley was dammed to create a small fishing lake in the 1950s. Now owned by the local wildlife trust, the open water and surrounding mire and scattered trees are managed as a nature reserve. The small wet woodland is dominated by willow, providing perching and nesting habitat for birds including linnet, reed bunting and song thrush. Jack snipe and water rail hide in the margins.

To reach the reserve, park in the lay-by next to the church (Access point) and follow the public footpath through the churchyard and past the farmhouse for about 400m. The site is wet throughout the year, its boardwalks only existing where otherwise deep wading would be required.

BAILEY EINON

LLANDRINDOD/LLANDRINDOD WELLS, POWYS   MAP REF 173

Ownership: Private: Radnorshire Wildlife Trust
Designations: SAC, SSSI
Area: 5ha

Forest type: broadleaved
Forest location: SO082615
Explorer Map: 200
Ease of access: Easy/Moderate

Access point: SO0847561198
///tend.quoted.openly
LD1 5SP

Bailey Einon is an ancient ash woodland with alder and hazel growing on the west bank of the Afon Leithon/River Ithon, close to Shaky Bridge. It is rich with bryophytes, invertebrates (all three species of cardinal beetle, and the snail-hunting beetle *Cychrus caraboides*), flowering plants (bluebell, dog's mercury, yellow archangel), birds (cuckoo, dipper, pied flycatcher, redstart, woodcock) and mammals (Daubenton's bat, otter). Due to the presence of ash dieback, which can make the trees prone to dropping branches, paths through the wood may be subject to closure. However, there is much to explore locally in this beautiful valley.

On the opposite bank is a large medieval field system and site of an abandoned village which once held a Royal Charter. The reconstructed twelfth-century church of Llanfihangel Cefnllys stands at its centre, accompanied by several 1,000-year-old yew trees. To the east is the hilltop castle of Cefnllys built by the powerful marcher lords during the same era, now offering wonderful views of Bailey Einon, the Leithon valley, and beyond.

Park in a lay-by near Shaky Bridge (Access point), where a stable wooden bridge has replaced the original nineteenth-century wire-hung and once aptly named bridge.

Llanfihangel Cefnllys church and Bailey Einon [173].

GRO WOOD

RHAEADR GWY/RHAYADER, POWYS　　　　　　　　　　MAP REF 172

Ownership: Private: Elan Valley Trust **Designations:** IPA, SAC, SM, SPA, SSSI	**Area:** 49ha **Forest type:** conifer **Forest location:** SN919627 **Explorer Map:** 200	**Ease of access:** Moderate **Access point:** SN9007461632 ///crowned.sprouts.centrally LD6 5HF

Stretching along the southern shore of Caban-coch reservoir, the mainly coniferous plantation presents an obvious opportunity for a scenic walk. Indeed, a 7km circular trail loops through and above the trees, providing wonderful views across the Elan Valley. Yet this wood holds hidden gems close to its heart. Scattered veteran oak and birch trees, which pre-date the conifers, host rich bryophyte communities including tree lungwort. Other rare lichens include *Chrysothrix chrysophthalma* and *Parmelietum laevigatae* growing on the tree trunks and branches.

At the extreme northern tip of the wood are the remains of the Nant-y-Gro dam (SN9219063480 | ///leathers.lectures.teeth), successfully breached in 1942 as part of the Royal Air Force's experiments with the dam-busting 'bouncing bombs' developed by Barnes Wallis. This historical feature is more easily reached from the Elan Visitor Centre (see Cnwch Wood [168]).

Birch and lichens.

Water-Break-its-Neck and Warren Wood [175].

WARREN WOOD

MAESYFED/NEW RADNOR, POWYS　　　　　　　　　　　　　　　　MAP REF 175

Ownership: Public: Natural Resources Wales **Designations:** **Area:** 65ha	**Forest type:** conifer **Forest location:** SO183608 **Explorer Map:** 200 **Ease of access:** Easy/Moderate	**Access point:** SO1943159338 ///bumping.grumbles.contoured LD8 2TN

Part of Radnor Forest [162], this small plantation is famous, not for its origins as a rabbit warren (farm), but as the site of one of the most spectacular waterfalls in Wales. The curiously named Water-Break-its-Neck can vary from a mere trickle in the summer to a roaring torrent in winter. A range of waymarked trails lead to the falls and beyond. Some of the larger conifers, including some impressive monkey puzzle trees, were planted by Victorians, who first popularised the waterfall spectacle.

Leave the A44 at the Access point and drive up the long track to reach a car park.

152　The Forest Guide: Wales

LLANHAYLOW WOOD

GLADESTRY, POWYS   MAP REF 176

Ownership: Private: Woodland Trust
Designations:
Area: 12ha

Forest type: mixed
Forest location: SO226565
Explorer Map: 201
Ease of access: Moderate

Access point: SO2246856215
///derailed.tallest.started
HR5 3NS

This was once an ancient woodland site but the broadleaved trees were felled during the 1950s and replaced with conifers, including Douglas fir, grand fir, Norway spruce, western hemlock and western red cedar. The new owner is now encouraging broadleaves to take over, although pockets and individual specimens of conifers remain, especially some towering specimens of Douglas fir near the Access point. The wood therefore has an attractive mix of tree species of different heights and ages. A network of informal paths can be followed around the wood. Near the Gilwern Brook these can be steep and involve sections with steps.

CORS Y LLYN

NEWBRIDGE-ON-WYE, POWYS   MAP REF 177

Ownership: Public: Natural Resources Wales
Designations: NNR, SSSI
Area: 8ha

Forest type: mixed
Forest location: SO014552
Explorer Map: 200
Ease of access: Easy

Access point: SO0163055676
///weeks.vegetable.skips
LD2 3RU

The diminutive nature of the birch and Scots pine trees on this National Nature Reserve masks their true age. The woodland is more than 100 years old, but the trees are stunted by the waterlogged peaty soils. The site is best known for its flower-rich meadows, but look for pied flycatcher, redstart and wood warbler in the woods. During the summer months, hobby hunt for dragonflies over the pond and open glades, while woodcock hide in the woods during the autumn. A short path leads to a circular boardwalk, making access relatively easy.

GARTH BANK

GARTH, POWYS   MAP REF 178

Ownership: Public: Natural Resources Wales
Designations: SSSI
Area: 73ha

Forest type: mixed
Forest location: SN946505
Explorer Map: 188
Ease of access: Moderate

Access point: SN9526151171
///skewing.cunning.revives
LD2 3LN

Garth Bank (301m) is the name of a prominent hill in the landscape, yet it is seldom visited by those from far afield. Its conifers have been extensively felled and replaced with a more diverse range of species. A network of forest rides and informal paths can be followed throughout the site. An old quarry at the north-western flanks of the hill (SN9463950889 | ///broth.binds.handed) is designated a geological SSSI. The rocks include those at the boundary between the Ordovician and Silurian periods, 433.8 million years ago. The older rocks hide fossil remains of brachiopods, a type of marine animal with two shells, which was very abundant at this time.

IRFON FOREST

LLANWRTYD, POWYS

MAP REF 179

Ownership: Public: Natural Resources Wales
Designations:
Area: 1,494ha

Forest type: conifer
Forest location: SN856507
Explorer Map: 187
Ease of access: Easy/Moderate

Access point: SN8565450759
///horseshoe.squabbles.recent
LD5 4TW

Irfon Forest lies north of Llanwrtyd, straddling the valleys of the Afon Cnyffiad and Afon Irfon. The forest is dominated by Sitka spruce (75 per cent), with smaller areas of Douglas fir, Norway spruce and lodgepole pine. Foresters are attempting to provide arboreal corridors for red squirrel, and replanting with more diverse conifer species (i.e. less Sitka spruce) to provide a good food source for the threatened mammal.

Informal trails within the forest are popular with mountain bikers. Given its large scale, there are many places where the forest can be accessed, including by walking or cycling from town. The main car park is at White Bridge (Access point) from where two waymarked trails can be followed alongside the Afon Irfon.

Red squirrel.

CWM BYDDOG

CLEIRWY/CLYRO, POWYS

MAP REF 180

Ownership: Private: Radnorshire Wildlife Trust
Designations: SM
Area: 4ha

Forest type: broadleaved
Forest location: SO217447
Explorer Map: OL13
Ease of access: Moderate

Access point: SO2150944727
///unfounded.rotations.splits
HR3 5SL

It is a wonder that this small but special nature reserve holds no conservation designations. The wooded dingle is packed with ancient and veteran trees, including the two majestic oaks, which frame the entrance to the reserve. The very rare (and protected) oak polypore is found here; its velvety brown fan-shaped fruiting body can be easily overlooked or confused with other bracket fungus species. The wood is managed specifically to provide suitable habitat for the nationally rare dormouse. A whole host of lichens, mosses, ferns and flowering plants thrive under the ancient trees, which in turn support a good range of woodland birds, including pied flycatcher.

Near the entrance, the motte and bailey of Castle Kinsey, dating from the medieval period, was constructed to take full advantage of the deep ravine of Cwm Byddog.

GARTH DINGLE

LLOWES, POWYS

MAP REF 181

Ownership: Private: Woodland Trust
Designations: SSSI
Area: 7ha

Forest type: broadleaved
Forest location: SO189421
Explorer Map: OL13
Ease of access: Moderate

Access point: SO1907541723
///community.sprain.shutting
HR3 5JG

Like nearby Cilcenni Dingle [182], the steep-sided wooded valley (the definition of a dingle) at Garth is special because it has lain largely undisturbed due to its inaccessibility. Sessile oak dominates the canopy, accompanied by rowan, silver birch, wild cherry and wych elm. Look for the bright orange and pink flowers of spindle, which is otherwise uncommon in this area.

Among the rich ground flora, enchanter's nightshade, herb-robert, early-purple and twayblade orchid thrive. Park in the lane and look for a public footpath sign (Access point), which starts between houses before leading north towards the dingle. Nearby Fron Wood (also owned by The Woodland Trust) can also be accessed from here.

CILCENNI DINGLE

LLOWES, POWYS

MAP REF 182

Ownership: Private: Woodland Trust
Designations: SSSI
Area: 17ha

Forest type: broadleaved
Forest location: SO173414
Explorer Map: OL13
Ease of access: Difficult

Access point: SO1732941224
///constants.hunt.packing
LD2 3UQ

This deeply incised wooded valley is the perfect example of a dingle. Its 1.5km-long precipitous slopes have protected much of the woodland from interference from humans, while all around the land has been farmed for centuries. Some of the accessible lower areas to the east were felled in the 1960s and planted with Douglas fir. The main canopy of the dingle is sessile oak, with an understorey of wych elm and hazel. Keep an eye out for the native black poplar. Dormice thrive here, as do many rare plants including herb-Paris, mudwort, twayblade and thin-spiked wood-sedge. Otter, grey wagtail and kingfisher live in the stream, along with white-clawed crayfish. Access within the wood is demanding, the slopes steep and often muddy, and everywhere the undergrowth is dense. Nearby Garth Dingle [181] is more accessible.

Wych elm.

GLASBURY CUTTING

Y CLAS-AR-WY/GLASBURY, POWYS MAP REF 183

| Ownership: Private: Wildlife Trust of South and West Wales
Designations:
Area: 2ha | Forest type: broadleaved
Forest location: SO183393
Explorer Map: OL13
Ease of access: Moderate | Access point: SO1870539593
///workroom.hypocrite.breath
HR3 5PT |

A finger of broadleaved trees makes up this small nature reserve growing along a disused railway line. Originally the route was a horse-drawn tramroad, constructed in 1818 to transport coal and wool, later converted to a steam railway. Eight years after the line was closed in 1962, the land was given to the local wildlife trust to create a nature reserve. Ash, silver birch and hazel have since naturalised, with carpets of primrose and other flowers lining the embankments. Signs of dormice feeding indicate that this special mammal is living here alongside common shrew and wood mouse. Access to the wood is a little hidden along the verge of the B4350, marked by a flight of steps climbing up onto the railway embankment (Access point). There is ample parking in a large lay-by 200m to the east, but beware the busy traffic and uneven verge.

BRYCHEINIOG FOREST

LLANEGLWYS, POWYS MAP REF 184

| Ownership: Private
Designations:
Area: 603ha
Forest type: broadleaved | Forest location: SO045391
Explorer Map: 188
Ease of access: Moderate | Access point: SO0569237843
///downcast.heads.auctioned
LD2 3BQ |

Within a decade of the Forestry Commission's formation in 1919 and its subsequent drive to afforest more of Britain aiming to increase homegrown timber production, it began to receive criticism for its lack of sensitivity to landscape and visual design. Created in the early 1930s, Brycheiniog Forest, originally Brecon Forest, was one of the first plantations laid out with more attention given to its appearance in the countryside. Its trees curve around a cwm of the Sgithwen Brook. Nestling at its foot, the hamlet of Llaneglwys was built to house foresters employed in creating and managing the forest. The village hall today is the old Nissen hut installed originally as a social centre for forest workers and local farmers. Several informal routes can be followed through the peaceful forest, usually in complete solitude other than with the company of a circling red kite or buzzard.

Red kite.

COED NANT BRÂN

LLANFIHANGEL NANT BRÂN, POWYS — MAP REF 185

Ownership: Private: Eco-explore Community Interest Company **Designations:** **Area:** 10ha	**Forest type:** broadleaved **Forest location:** SN938349 **Explorer Map:** 188 **Ease of access:** Moderate	**Access point:** SN9382534959 ///modest.grounding.oatmeal LD3 9NA

With close links to local universities and the wildlife trust, the owners conduct scientific research in this quiet valley. More than 60 nest boxes have been installed in the wood as part of an extensive monitoring programme, used for student teaching and to study bird behaviour. For instance, the date of the first egg laid each year helps researchers track the effects of climate change and to study bird breeding success. Tits (blue, great and long-tailed), nuthatch, treecreeper and pied flycatcher are commonly seen among the sessile oak trees and in the understorey of hazel and blackthorn.

An oak tree with nest box.

PWLL-Y-WRACH

TALGARTH, POWYS — MAP REF 186

Ownership: Private: Wildlife Trust of South and West Wales **Designations:** NP, SSSI **Area:** 17ha	**Forest type:** broadleaved **Forest location:** SO164327 **Explorer Map:** OL13 **Ease of access:** Easy/Moderate	**Access point:** SO1619032794 ///foggy.require.veto LD3 0DT

Meaning 'Witches' Pool Valley' in English, this narrow coombe is lined with ash, sessile oak and field maple, with hawthorn, rowan and hazel understorey. Honeysuckle twines around many of the trees and its heady scent fills the air on a summer evening. It is a favourite food for dormice, which live among the hazel trees. Otter, kingfisher and dipper frequent the Afon Ennig, which tumbles over several attractive waterfalls. In shady areas, look for two rare and special parasitic plants, both of which do not photosynthesise but rely on nutrients stolen from their hosts: the bird's nest orchid and toothwort. Given their complete absence of green colour and sickly appearance, it might seem they survive only thanks to black magic! From a small car park (Access point), a level path, which is suitable for wheelchairs, leads into the heart of the reserve to a bench, yet the main waterfall lies beyond and is reached via an undulating and often-muddy path.

A red kite soars over Cwm Sere [189].

HALFWAY FOREST

PONTSENNI/SENNYBRIDGE, POWYS   MAP REF 187

Ownership: Public: Natural Resources Wales **Designations:** SAC, SSSI **Area:** 389ha	**Forest type:** coniferous **Forest location:** SN843347 **Explorer Map:** OL12 **Ease of access:** Moderate	**Access point:** SN8357433001 ///increases.spokes.tightrope SA20 0SE

Halfway Forest borders the Pontsenni/Sennybridge military training area, like its larger neighbour Crychan Forest [206] in the South-West region. Straddling Carmarthenshire and Powys, this plantation of Sitka spruce grows up to 420m altitude. Waymarked routes, including a section of the Epynt Way, pass through forest and treeless upland (including an upland SAC/SSSI). An option to walk to the Fedw viewpoint (SN8610031771 | /// rationing.cover.fake) along the Epynt Way guides the visitor between two of the curious rectangular plantations (lying between marker posts 542 and 543) described in the Crychan Forest entry.

GLASFYNYDD FOREST

TRECASTELL/TRECASTLE, POWYS   MAP REF 188

Ownership: Public: Natural Resources Wales **Designations:** NP **Area:** 1,351ha	**Forest type:** conifer **Forest location:** SN825275 **Explorer Map:** OL12 **Ease of access:** Easy/Moderate	**Access point:** SN8199727136 ///blip.sweated.bugs LD3 8YE

Glasfynydd Forest comprises three substantial conifer blocks surrounding Usk Reservoir at the northern tip of the Bannau Brycheiniog/Brecon Beacons National Park. Its remote location within an International Dark Sky Reserve makes this one of the best areas in Britain for astronomical gazing. The forest is being diversified to make it more resilient to pests and pathogens, improve amenity and enhance biodiversity. Look for goosander on the reservoir, and nightjar or woodcock along forest rides. Crossbill and goldcrest share the tree canopies with pine marten. A 9km-long waymarked trail circumnavigates the reservoir, which on the north side lies within Carmarthenshire (South-West region). Enjoy good views to Y Mynyddoedd Duon/Black Mountains.

CWM SERE

ABERHONDDU/BRECON, POWYS   MAP REF 189

Ownership: Private: National Trust **Designations:** NP, SSSI **Area:** 57ha **Forest type:** broadleaved	**Forest location:** SO026232 **Explorer Map:** OL12 **Ease of access:** Moderate/Difficult	**Access point:** SO0247624864 ///attracts.crumple.alone LD3 8LE

Every year, countless thousands of walkers look down on the distant trees of Cwm Sere from the peaks of Pen y Fan and Cribyn, but very few deviate from the obvious climber's route to follow instead the wandering sheep tracks to explore the sessile oak trees and hidden waterfalls of this beautiful cwm.

From the National Trust car park (charge), start by climbing up a well-trodden route south for 300m before leaving the crowds behind to contour around Allt Ddu and head towards the wood. Look for raven and red kite circling overhead, and grey wagtails feeding from the bright waters of Nant Sere.

ALLT YR ESGAIR WOODS

LLANSANTFFRAED, POWYS MAP REF 190

Ownership: Private
Designations: NP, SM
Area: 35ha
Forest type: mixed

Forest location: SO124244
Explorer Map: OL13
Ease of access: Moderate

Access point: SO1294322704
///truth.wildfires.flipping
LD3 7JQ

Climb this prominent hill on the east side of the River Usk to enjoy spectacular views across the Welsh countryside. Far below, the contorted meanders of the river threads across the floodplain and patches of alluvial forest. Immediately beyond stretches a patchwork of pastures and ancient hedgerows, and above them the dramatic skyline of the Bannau Brycheiniog/Brecon Beacons. In the opposite direction lie the Y Mynyddoedd Duon/Black Mountains. With such vistas, it is no wonder that a hill fort was established here in the Iron Age. Its name means wooded ridge, and today plantations of young trees line the sides of the path, which can be steep and sometimes muddy as it heads up the hill from the Access point.

CRAIG CERRIG GLEISIAD

ABERHONDDU/BRECON, POWYS MAP REF 191

Ownership: Public: Natural Resources Wales
Designations: IPA, NNR, NP, SAC, SSSI

Area: 11ha
Forest type: broadleaved
Forest location: SN963218
Explorer Map: OL12

Ease of access: Difficult
Access point: SN9713422242
///chew.comply.fame
LD3 8NL

This spectacular curved valley was gouged from the mountains by a glacier during the last Ice Age, creating the dramatic cliffs at its head and a sheltered corrie below. Luckily for visitors, relatively few people are attracted to this wonderful place, lured instead by the barren high peaks of the Bannau Brycheiniog/Brecon Beacons or the dramatic scenery of waterfall country (e.g. Coed Mellte [199]).

The view from Allt yr Esgair [190].

Many rare plants grow among the boulders and steep crags. At higher altitudes and north-facing slopes, arctic-alpine wildflowers prosper, including mossy and purple saxifrage, green spleenwort and serrated wintergreen. This is the southernmost location for these plants, which do not occur again before the Alps. Besides the sparse woodland of the corrie floor, scattered trees grow on the near-vertical crags high above. Among the hawthorn, rowan and occasional stunted ash trees are scattered whitebeam trees. Keen botanists can hunt for some rare micro-species of whitebeam found on other north-facing slopes in the Bannau Brycheiniog. Similar sites include the NNR of Craig y Cilau [193] and Cwm Claisfer [192]. Listen for the distinctive cry of peregrine falcon, which nests nearby.

Humans once took full advantage of both the natural shelter provided by the crags, and the commanding views across surrounding wooded countryside. Prehistoric hut circles are present nearby, as are the remains of a deserted settlement. Since then, hill farming with sheep has removed most of the trees. Patches of bluebells can be seen under the bracken in spring, providing proof that much more of the now-treeless hill was once covered by trees. One day in the future, improved policies and changes in land management practice may help recover biodiversity on this site. Currently however, the plethora of conservation designations seems incongruous.

Walk from a lay-by on the A470 (Access point), following a distinct path south-west and uphill towards the crags. Visitors should be well equipped for hillwalking and should not attempt to explore steeper sections of the crags. The Beacons Way follows the ridge above the trees, taking in the summit of Craig Cerrig Gleisiad (629m).

Scattered trees and sparse woodland at Craig Cerrig Gleisiad [191].

Central-East **161**

CWM CLAISFER

LLANGYNIDR, POWYS MAP REF 192

Ownership: Public: Natural Resources Wales
Designations: IPA, NP
Area: 31ha
Forest type: mixed
Forest location: SO145168
Explorer Map: OL13
Ease of access: Difficult
Access point: SO1500517376
///snapping.represent.club
NP8 1LL

Cwm Claisfer lies at the head of the valley south of Llangynidr, offering dramatic views of the Usk Valley and the distant Y Mynyddoedd Duon/Black Mountains. It is another site within the Cliffs of the Bannau Brycheiniog/Brecon Beacons National Park IPA, and like Craig Cerrig Gleisiad [191] and Craig y Cilau [193] is home to a spectacular range of rare plants. A small number (c.27) of lesser whitebeam trees grow here, significantly fewer than their stronghold at Craig y Cilau, where more than 700 specimens have been counted. The richest biodiversity is found immediately below the limestone escarpment, where ash and alder dominate. The ground is covered in boulders, and like the trees, carpeted in mosses, lichens and liverworts. Look and listen for spotted flycatcher, tree pipit and wood warbler. Further downslope lies a conifer plantation.

There are two main options to reach the wood. From the Access point, a small car park on a sharp bend of the B4560, either follow a track to the plantation or follow a path heading up towards the crags. The alternative (SO1570217087 | ///transit.breeze.renew) a little further up the B4560 provides a route with less climbing, contouring around the hill below a quarry to reach the foot of the crags, although the path is indistinct.

CRAIG Y CILAU

LLANGATWG/LLANGATTOCK, POWYS MAP REF 193

Ownership: Public: Natural Resources Wales
Designations: IPA, NNR, NP, SAC, SSSI
Area: 14ha
Forest type: broadleaved
Forest location: SO184164
Explorer Map: OL13
Ease of access: Moderate/Difficult
Access point: SO1852416893
///rely.awakes.unpainted
NP8 1PZ

As at similar north-facing cwms in the Bannau Brycheiniog/Brecon Beacons (e.g. Craig Cerrig Gleisiad [191], Cwm Claisfer [192]), the steep and inaccessible cliffs and scree of this site provide a haven for rare wildlife. Its lower slopes are lime-rich grassland, which is a relatively rare habitat in Wales, while impressive limestone cliffs tower above, creating a natural amphitheatre. The cliffs were formed by quarrying during the eighteenth and nineteenth centuries, the limestone taken by horse-drawn tram to the ironworks near Brynmawr.

The site is rich with more than 250 species of flowering plants. Birdlife abounds (look for lesser spotted woodpecker, redstart and peregrine falcon),

English whitebeam at Craig y Cilau [193].

while caves in the cliffs support several bat species, notably lesser horseshoe. Among the ash, birch, oak and rowan trees are scattered several micro-species of whitebeam. Notably, this is one of the few sites anywhere in the world where thin-leaved whitebeam and lesser (also known as 'least') whitebeam grow.

Park on the verge of the unclassified road just above the cattle grid and follow the path (Access point), which also forms part of the Cambrian Way, southwards towards the cliffs.

Looking east from Craig y Cilau [193].

TALYBONT FOREST

TALYBONT-ON-USK, POWYS MAP REF 194

Ownership: Public: Natural Resources Wales
Designations: LNR, NP
Area: 791ha

Forest type: conifer
Forest location: SO060173
Explorer Map: OL12
Ease of access: Easy/Moderate

Access point: SO0626717047
///attending.barbarian.skinning
LD3 7YT

At many points among this large conifer plantation surrounding Talybont Reservoir, pleasant views can be enjoyed across the water through stands of towering spruce trees. Both the Beacons Way and the Taff Trail pass through the forest, east of the reservoir. At its western end, the Afon Caerfanell tumbles over a series of pretty waterfalls reached via a path from a car park (Access point) near Pont Blaen-y-glyn.

CRAIG Y RHIWARTH

GLYNTAWE, POWYS MAP REF 197

Ownership: Private: Wildlife Trust of South and West Wales
Designations: IPA, NP, SSSI
Area: 22ha

Forest type: broadleaved
Forest location: SN843157
Explorer Map: OL12
Ease of access: Moderate

Access point: SN8396715537
///whizzing.risen.reserved
SA9 1GL

This pretty woodland with a rich ground flora is managed as a nature reserve. Sadly, its ash trees are terribly afflicted by dieback, exposing the hazel understorey. It will be interesting to observe in future if the hazel will form the main canopy, or whether another species takes over. The woodland is on a steep slope and is mostly quite inaccessible, but a reasonably level path passes through the fringes of the wood. Look for harebell, lily of the valley, wild thyme and wood spurge. Dipper and kingfisher dart along the fast-flowing Afon Tawe. A small number (c.29 trees) of thin-leaved whitebeam grow here, one of only two sites in the Bannau Brycheiniog/Brecon Beacons (see also Craig y Cilau [193]).

The easiest Access point is to park (charge) at the nearby Craig y Nos Country Park and then to walk north along the eastern bank of the river through a mixed woodland of Scots pine and broadleaves. The entrance to the reserve (Access point) is near a wooden pedestrian bridge. An alternative is a nearby lay-by (SN8431415716 | ///shuffles.repeating.flicked), although this requires a longer walk to reach the woodland. Allt Rhongyr nature reserve can also be reached from the country park.

Ash dieback at Craig y Rhiwarth [197].

TAF FECHAN FOREST

PONTSTICILL, POWYS MAP REF 195

Ownership: Public: Natural Resources Wales **Designations:** NP, SM **Area:** 690ha	**Forest type:** conifer **Forest location:** SO042166 **Explorer Map:** OL12 **Ease of access:** Moderate	**Access point:** SO0484016265 ///resettle.bake.dairies CF48 2UT

Taf Fechan Forest sites immediately below the high peaks of the Bannau Brycheiniog/Brecon Beacons, its conifers extending down a steep-sided valley before enclosing Pentwyn and Pontsticill reservoirs. The forest previously included a substantial quantity of larch, whose golden hues in autumn looked spectacular when reflected in the surfaces of the reservoirs, but sadly most have been felled due to the water mould *Phytophthora ramorum*.

The ruins of a seventeenth-century farmstead lie within the trees. Look for signs of otter near the water, and the bright flash of kingfisher and grey wagtail. Goldcrest and crossbill thrive in the canopies of the spruce trees.

A river walk can be followed from the Owl's Grove car park (Access point), which although not strenuous can be muddy. Torpantau railway holt is nearby, the end of the line for the Brecon Mountain Railway, offering an interesting alternative means of arrival. Another car park at Pont Cwmyfedwen (SO0425216409 | ///conqueror.effort.stars) can be found further upstream the Taf Fechan river, joining the same trail. Further upstream still, a car park at Neuadd (SO0366817064 | ///perplexed.flamingo.scoots) not only lies closer to the attractive Blaen y Glyn falls, but is used by walkers attempting the famous Horseshoe Trail, which includes the peaks of Cribyn (795m) and Fan y Big (717m). Sustrans Route 8 passes through the forest, as does the Taff Trail.

PLAS-Y-GORS

YSTRADFELLTE, POWYS MAP REF 196

Ownership: Public: Natural Resources Wales **Designations:** NP, SM **Area:** 185ha	**Forest type:** conifer **Forest location:** SN923162 **Explorer Map:** OL12 **Ease of access:** Easy/Moderate	**Access point:** SN9272516462 ///passively.august.cocktail CF44 9JE

Situated high on the moorland slopes of the Bannau Brycheiniog/Brecon Beacons, near the river sources of waterfall country (see Coed Mellte [199] and Coed Nedd [200]), the twentieth-century planners of this isolated coniferous plantation were perhaps ignorant of the site's historical significance. The remains of a large rectangular (390×240m) Roman marching camp were later discovered under the trees (now felled). It was once used as a temporary camp by legions as they marched westward between AD 74 and AD 77. The camp is also dissected by a major Roman road known as Sarn Helen, whose 260km route links Aberconwy with Carmarthen. Nearby, just south of trees, is the menhir known as Maen Madoc (SN9186915783 | ///nutty.congas.splits), which bears a vertical inscription in Latin on its south-west side, which translates as 'Dervacus, Son of Justus. He lies here'. From the car park (Access point), follow the Beacons Way south-west.

WERN PLEMYS

YSTRADGYNLAIS, POWYS   MAP REF 198

Ownership: Private: Wildlife Trust of South and West Wales **Designations:** **Area:** 5ha	**Forest type:** broadleaved **Forest location:** SN788092 **Explorer Map:** 165 **Ease of access:** Easy/Moderate	**Access point:** SN7861709476 ///reviews.wordplay.screen SA9 1ES

Ystradgynlais town thrived with burgeoning coal, iron and lime industries. Much of the industrial landscape has been cleared and landscaped, including the site of the former Diamond Colliery, which now comprises Diamond Park. The nature reserve of Wern Plemys lies beyond the park (Access point).

Areas of mixed broadleaved woodland, often dominated by ash with willow, surround wet meadows rich with wildflowers and grazed by ponies. Sustrans Route 43 passes to the south of the reserve along the trackbed of an old industrial railway line.

COED MELLTE

PONTNEDDFECHAN, POWYS   MAP REF 199

Ownership: Public: Natural Resources Wales and Bannau Brycheiniog National Park Authority **Designations:** IPA, NP, SAC, SM, SSSI	**Area:** 219ha **Forest type:** temperate rainforest **Forest location:** SN924106 **Explorer Map:** OL12 **Ease of access:** Moderate	**Access point:** SN9112207932 ///hobbit.flamingo.gamer SA11 5NB

The Afonydd (rivers) Mellte, Hepste, Pyrddin and Nedd-fechan/Neath collectively form 'waterfall country' in the south-west of the Brecon Beacons National Park. Their waters tumble off the mountains in a series of dramatic waterfalls, carved through steep and tree-lined gorges before joining the Afon Nedd (River Neath).

Coed Mellte lies at the heart of waterfall country, featuring no fewer than nine major cascades. Together with neighbouring Coed Nedd [200], it is part of the Coedydd Nedd a Mellte SAC, recognised as one of the largest and most diverse areas of sessile oak woodland and temperate rainforest in Wales. Although dominated by oak, ash is also present, with birch, field maple, rowan and small-leaved lime. The woods are rich with mosses (e.g. flagellate feathermoss, Irish daltonia), lichens (e.g. autumn flapwort, straggling pouchwort, pearl pouncewort, western earwort) and ferns (e.g. green spleenwort and Tunbridge filmy fern).

A car park (charge) near Craig y Ddinas/Dinas Rock is the Access point given, as it offers the easiest access. Along with some lower sections of the gorge, it lies within Rhondda Cynon Taf/South Wales Central but the main body of the wood is within Powys. The site is popular with walkers, climbers and gorge scramblers and therefore can become very busy. An accessible trail (600m) takes in the Sychryd cascades. A more strenuous (6km) trail leads to Sgwd yr Eira falls, allowing exploration of the nineteenth-century Glyn-nedd/Glynneath gunpowder works (one of only two in Wales) including ruins of a pellet house, magazine, tramroad and leat. It also passes near the Iron Age hill fort of Craig y Ddinas occupying the sheer cliffs overlooking the Afonydd Mellte and Sychryd.

Many walkers undertake the 9km (Moderate) 'four waterfalls walk' which visits Sgwd Clun-Gwyn, Sgwd Isaf Clun-Gwyn, Sgwd y Pannwr and Sgwd yr Eira. The route starts from one of two car parks (charge) further up the valley, either at Gwaun Hepste (SN9356912362 | ///originals.reclusive.surfaces) or the smaller car park at Cwm Porth (SN9285012433 | ///tramps.harshest.bedrock). Both quickly reach capacity in high season. A good alternative is to take advantage of a free shuttle bus from Glyn-nedd/Glynneath which drops passengers off at both car parks.

COED NEDD

PONTNEDDFECHAN, POWYS

MAP REF 200

Ownership: Public: Natural Resources Wales and Brecon Beacons National Park Authority **Designations:** NP, SAC, SM, SSSI	**Area:** 163ha **Forest type:** temperate rainforest **Forest location:** SN901095 **Explorer Map:** OL12	**Ease of access:** Moderate **Access point:** SN9007607653 ///coaster.warmers.sharpened SA11 5NR

Coed Nedd forms part of one of the largest and most diverse areas of sessile oak woodland and temperate rainforest in Wales. Together with Coed Mellte [199], it lies at the heart of waterfall country, yet unlike its neighbour it suffers from much less footfall. Mosses, liverworts, lichens and ferns cloak every surface with emerald green. Grey wagtail brighten the tumbling waters where kingfisher and otter hunt.

The Elidir Trail (8km return), named after a runaway trainee monk supposedly taken in by the fairies of the cwm, can be followed between Pontneddfechan and Pont Melin-fach. The initial section follows an old tramway past quarries and remains of mine buildings, and is easy going until the first bridge. A short diversion up the Afon Pyrddin tributary leads to the most spectacular of the four main waterfalls on the trail, named Sgwd Gwladus. Beyond the bridge it is a little more demanding, involving steps and short gradients, but the views are spectacular, taking in the multiple falls along the steep wooded gorge of the Afon Nedd. A second bridge crosses the Afon Nedd, allowing exploration of the entrance to an old silica mine on the east side of the river.

Access to the wood and the start of the Elidir Trail is behind the pub at the junction of Pontneathvaughan Road with the High Street, with free parking available in a small car park or nearby along the main road.

Sgwd Gwladus waterfall at Coed Nedd [200].

The South-West is a large region, second only to Central-East, incorporating the counties of Pembrokeshire, Carmarthenshire, Swansea and Neath Port Talbot. The entire Pembrokeshire National Park falls within its western boundary, while part of the Bannau Brycheiniog/Brecon Beacons National Park is within its extreme eastern end. About 14 per cent of the region is covered with trees, falling just below the average for Wales (15 per cent). Many spectacular coastal woodlands can be explored, such as Aber Mawr [217], Bathesland Wood [231] and Castell Llansteffan [237], plus several sites on the Gower Peninsula such as Coed Nicholaston [262], Oxwich Wood [263], and Pennard Cliff and Northill Woods [261]. Away from the coast, the

Oxwich Bay and Oxwich Wood [263].

landscape of Pembrokeshire and Carmarthenshire has a distinctive beauty, characterised by thickly hedged pasture fields with woodlands limited to steep-sided valleys. Ffynone and Cilgwyn Woods [210] and Coed Pencastell [223] are characteristic examples of the region's dingles. Some temperate rainforest sites exist at the eastern border with Central-East, such as Blaenant y Gwyddyl [239] and Gwenffrwd-Dinas [201], in stark contrast to Kilvey Hill [255] overlooking Abertawe/Swansea city. The region has a rich mining history, from gold at Dolaucothi [208], to coal at Craig Gwladus [248] and Cwm Clydach [246]. Meanwhile, the valley known as Parc le Breos Cwm/Coed y Parc [259] has a much older history waiting to be explored.

SOUTH-WEST

PEMBROKESHIRE, CARMARTHENSHIRE, SWANSEA, NEATH PORT TALBOT

SITES 201–263

GWENFFRWD-DINAS

RHANDIRMWYN, CARMARTHENSHIRE   MAP REF 201

Ownership: Private: RSPB	**Forest location:** SN780465	**Access point:** SN7876647102
Designations: IPA, SAC, SPA, SSSI	**Explorer Map:** 187	///saga.adjust.slanting
Area: 36ha	**Ease of access:** Moderate/Difficult	SA20 0NW
Forest type: temperate rainforest		

At the confluence of the Afon Pysgotwr with the larger Afon Tywi/River Towy, the wooded hill of Dinas rises steeply above the roaring waters of the rocky gorge. Its sessile oak trees and boulders are clad in green carpets of mosses, lichens and liverworts. Like nearby Allt Rhyd y Groes [202], this is a wonderful site for woodland birds, including marsh tit, pied flycatcher, redstart, tree pipit and wood warbler. Bluebells bloom spectacularly here in late spring.

From the car park (Access point), a circular trail can be followed anticlockwise around the hill. It begins with an easy section of boardwalk through alder carr woodland, but soon becomes very rough, with steep sections requiring some scrambling over rocks as it follows the tumbling waters of the Afon Tywi/River Towy.

Near the top of the hill is a cave reputedly used by the 'Welsh Robin Hood', Twm Siôn Cati, to avoid capture by the sheriff of Carmarthen. Born c.1530 at Tregaron, the thief with a reputation for canniness while eschewing violence robbed the rich but was less known for giving to the poor!

Approximately halfway round, a short but steep diversion from the circular route climbs the hill to reach the cave (SN7804446834 | ///rockets.horns.bloodshot) and includes sections with steps. The open-roofed cave is very atmospheric, its surfaces covered with carved graffiti left by generations of visitors.

The second half of the circular route is slightly easier going, and indeed if the cave is of main interest, an alternative car park (SN7822246356 | ///airliners.affords.scribble) offers a better out-and-back option. The section between the two car parks near the road is easiest of all; its wide-spaced sessile oak trees offering ample opportunity for spotting iconic woodland birds.

The boardwalk at Gwenffrwd-Dinas [201].

ALLT RHYD Y GROES

RHANDIRMWYN, CARMARTHENSHIRE MAP REF 202

Ownership: Public: Natural Resources Wales **Designations:** IPA, NNR, SAC, SPA, SSSI	**Area:** 73ha **Forest type:** temperate rainforest **Forest location:** SN760484 **Explorer Map:** 187	**Ease of access:** Difficult **Access point:** SN7754947085 ///sushi.providing.canyons SA20 0PL

Sessile oak trees with rowan, hazel and alder clad this dramatic gorge, supporting a wealth of wildlife. The scenery meanwhile is among the best-kept secrets in Wales. The vertical crags of Craig Pysgotwr dominate the north side of the river, providing nesting sites for peregrine and raven. The very steep wooded slopes to the south once harboured the last remaining pairs of red kite in Britain before the successful breeding effort of the 1990s led to their spectacular return across the country. Watch for merlin and meadow pipit on the open moorland, dipper and kingfisher near the rivers, and among the trees, pied flycatcher, redstart, treecreeper, wood warbler and woodpeckers.

The site is rich with bryophytes, including the rare Holt's mouse-tail moss, which grows only on boulders and tree roots regularly sprayed by fast-flowing water. Bluebells bloom spectacularly in May, while other plants to spot include devil's-bit scabious, great burnet, pignut and wood anemone.

The Access point is a track that begins next to a wooden bridge. There is room to park a car just across the bridge before returning on foot to follow the track upstream, keeping the Afon Pysgotwr on the right-hand side. Gwenffrwd-Dinas [201] lies nearby.

Allt Rhyd y Groes [202].

NANT MELIN

PUMSAINT, CARMARTHENSHIRE MAP REF 203

Ownership: Private: Wildlife Trust of South and West Wales **Designations:** IPA, SAC, SPA, SSSI **Area:** 34ha	**Forest type:** broadleaved **Forest location:** SN729466 **Explorer Map:** 187 **Ease of access:** Moderate	**Access point:** SN7321346558 ///thumbnail.daily.scrap SA19 8YD

This ancient upland sessile oak woodland is a small but perfect example of the much larger Cwm Doethie – Mynydd Mallaen SAC, which covers more than 4,000ha of the southern Mynyddoedd Cambrian/Cambrian Mountains, including Allt Rhyd y Groes [202]. The area west of the tumbling stream of Nant Melin is a nature reserve (2.4ha) managed by the local wildlife trust, the remainder to the east, privately owned.

The oak is accompanied by alder, birch, hazel and rowan. Royal fern grows in damp and shady areas, and the tree trunks and branches support multitudes of bryophytes. Look for both native species of flycatcher and wood warbler, and listen for cuckoo in the spring. A small lay-by (Access point) provides space for a single car near a track, which climbs into the wood.

FOREST WOOD

CILGERRAN, PEMBROKESHIRE

MAP REF 205

Ownership: Private **Designations:** NNR, SSSI **Area:** 18ha **Forest type:** broadleaved	**Forest location:** SN189440 **Explorer Map:** OL35 **Ease of access:** Easy/Moderate	**Access point:** SN1979142955 ///corrupted.wades.utter SA43 2SS

For centuries, timber was cut from the woods on either side of the lower reaches of the Afon Teifi to help build and repair the great Norman castle at Cilgerran. On the opposite bank lies Coedmore, meaning 'great wood' in Old Welsh, which lends its name to a designated NNR comprising woodland on both banks of the steep-sided valley. Access to Coedmore wood is limited except at Cwm Du where a lay-by (SN1961544512 | ///repaid.requests.portfolio) provides space for a couple of cars, and a short path can be followed down to the river.

Alternatively, a picturesque walk can be enjoyed along the left bank through Forest Wood, which also falls under the Coedmore NNR. A sawmill once operated in Forest Wood before slate mining became a more profitable enterprise. A wooden wharf was built on the riverbank to help load the slate onto ships but is now long gone. The old quarries are covered in scrub woodland and lie just outside the NNR. From the lower car park in town (Access point), walk north-west along the riverside path, enjoying glimpses through the trees to the castle high above. Forest Wood lies beyond the castle, after negotiating a shallow ford across the Afon Plysgog.

The woods are dominated by sessile oak, with occasional ash, small-leaved lime, wild service and wych elm. The Afon Teifi, designated as an SAC, is a haven for otter, which feast on the sea trout for which this river is well known by anglers. Pied and spotted flycatchers thrive on the abundant invertebrates, as do several bat species. The moist and shady environment supports a huge diversity of mosses, lichens and ferns.

The Teifi coracle was used for hundreds of years to fish on the river with nets, often by a pair of fishers working in separate coracles. Made from clefts of willow and a carry strap of twisted hazel, the small boats were originally covered in animal skin. An annual coracle race is held here every year.

The ruins of Cilgerran Castle above Forest Wood [205]. The riverside path is just visible.

CWM RHAEADR

RHANDIRMWYN, CARMARTHENSHIRE MAP REF 204

Ownership: Public: Natural Resources Wales **Designations:** **Area:** 142ha	**Forest type:** conifer **Forest location:** SN762430 **Explorer Map:** 187 **Ease of access:** Easy/Moderate	**Access point:** SN7651642243 ///hitters.recording.gymnasium SA20 0TL

Giant Douglas fir trees thrive in this sheltered cwm in the Tywi Valley. The Nant Rhaeadr tumbles dramatically from the crags at the head of the cwm, creating the highest waterfall in the county. Reaching the waterfall requires some significant ascent and the navigation of uneven paths, but an accessible trail can also be enjoyed from the car park (Access point), which includes a small pond with a section of boardwalk. The forest is popular with mountain bikers, who enjoy the thrill of a high-quality red-graded descent.

CRYCHAN FOREST

LLANWRTYD WELLS, CARMARTHENSHIRE — MAP REF 206

Ownership: Public: Natural Resources Wales **Designations:** SAC, SSSI **Area:** 2,047ha	**Forest type:** mixed **Forest location:** SN845402 **Explorer Map:** 187 **Ease of access:** Easy/Moderate	**Access point:** SN8489940997 ///overpaid.humans.recline SA20 0YU

North of the Bannau Brycheiniog/Brecon Beacons and south of the Mynyddoedd Cambrian/Cambrian Mountains lies a curious collection of small blocks of forest scattered across the otherwise treeless uplands. Seemingly haphazard and running against the grain of the landscape, the rectangular plantations were planted soon after the area was designated with a very specific purpose. They lie within the Pontsenni/Sennybridge military training area, also known as Mynydd Epynt. Hundreds of local people were forcibly evicted from their homes and farmsteads in 1940 to make way for military activities, and civilians have been excluded from the area ever since.

It is not permissible to explore the small plantations, but nearby Crychan Forest, which lies to the north-west of the military area, provides a great choice of waymarked routes for walkers, cyclists and equestrians. Recently a 65km-long recreational route, the Epynt Way, has been created by the Ministry of Defence to circumnavigate the area, while also offering a number of shorter circular routes. There are four main entry points into the forest, three of which also link with the Epynt Way, including at Brynffo (Access point). Another is at Halfway Forest [187](Central-East region).

Goshawks hunt among the vast stands of conifers, and honey-buzzard may be seen circling overhead in summer. A chance encounter with an abandoned farmstead and its old stone walls offers a reminder of the recent past.

A local community group, the Crychan Forest Association, supports the owners in promoting forest recreation, offering detailed guidance for visitors on a dedicated website.

The distinctive blocks of conifers at Mynydd Epynt seen from Fforest Fawr and Cray Reservoir, 12km to the south.

CAIO FOREST

PUMSAINT, CARMARTHENSHIRE   MAP REF 207

Ownership: Public: Natural Resources Wales **Designations:** **Area:** 1,069ha **Forest type:** conifer	**Forest location:** SN698409 **Explorer Map:** 187 **Ease of access:** Easy/Moderate/Difficult	**Access point:** SN6795240518 ///sobered.makeup.outboard SA19 8RF

High above the historic village of Caio, the stands of Sitka spruce and occasional broadleaves of this forest blanket the foothills of the Mynydd Mallaen plateau. Three waymarked trails can be followed from the car park (Access point), which is reached by following an unclassified road that becomes a track for the last 300m.

Directly west of the Access point on the open hill are the remains of the Annell Aqueduct and Leat. These historic artefacts once carried water to the Roman gold mines at nearby Dolaucothi [208]. It is clear from the contours of the hillside that the leat would have continued through the area now afforested to its confluence with the Afon Annell somewhere at the northern tip of the modern-day forest. The water was collected in tanks before being released as small floods to wash soil away from the bedrock. The technique, known as 'hushing', helped reveal the veins in the rock, which could then be extracted.

South-West 175

Mining infrastructure at Dolaucothi [208].

DOLAUCOTHI

PUMSAINT, CARMARTHENSHIRE — MAP REF 208

Ownership: Private: National Trust **Designations:** SSSI **Area:** 19ha **Forest type:** temperate rainforest	**Forest location:** SN665401 **Explorer Map:** 186 **Ease of access:** Easy/Moderate	**Access point:** SN6624340333 ///shuts.proof.rationing SA19 8US

The ancient mines at Dolaucothi are the only confirmed Roman source of British gold from the first and early second centuries AD. Water was cleverly drawn from the hills above, including the area now covered by trees at Caio Forest [207], to help reveal veins of quartz in the shale rocks, which contain tiny fragments (more dust than nuggets) of the precious metal. The mine was reopened in the 1930s but ultimately proved unprofitable.

Today, some of the mine adits provide a hibernaculum and roosting site for bats, including barbastelle, brown-eared, greater horseshoe, natterer's, noctule and pipistrelle (common and soprano). The surrounding area, which at one time would have been an industrial landscape, is now fully reclaimed by nature. Ancient sessile oak trees support myriad invertebrates that in turn provide food for the bats and many woodland birds, including pied flycatcher, redstart and treecreeper. The temperate rainforest environment provides perfect conditions for a rich diversity of lichens, mosses and ferns, including the hay-scented buckler fern.

The goldmines are operated as a visitor attraction for part of the year, but the car park (Access point) for woodland walks, which is found opposite, is open all year. It is possible to walk around the main mining areas, or to explore the tranquil river valley.

PENGELLI FOREST

FELINDRE FARCHOG, PEMBROKESHIRE — MAP REF 209

Ownership: Private: Wildlife Trust of South and West Wales
Designations: NNR, NP, SAC, SSSI
Area: 66ha
Forest type: temperate rainforest
Forest location: SN132391
Explorer Map: OL35
Ease of access: Easy/Moderate
Access point: SN1226139585
///parts.stance.frosted
SA41 3PX

A beautiful field gate and kissing gate, made from cleft (i.e. split not sawn) local timber, is a sure sign that traditional crafts are nurtured at Pengelli Forest. Indeed, the forest is home to the Coppicewood College (an educational charity), which teaches students traditional woodland management and green woodworking skills. It also welcomes volunteers.

Together with neighbouring Pant-teg Wood, Pengelli Forest is the largest area of ancient oak woodland in the region and is protected by several important conservation designations.

Next to the main entrance (Access point), where there is a small lay-by, the sessile oak and birch trees of Pant-teg Wood grow on a steep hillside with acid soils. Hazel and holly form the understorey, with common cow wheat and bilberry.

Pengelli Forest lies beyond, and is readily explored via a network of tracks, though these can be steep and rutted in places. Its trees grow on poorly drained boulder clay, which means that its habitat differs from Pant-teg Wood. Much of the main canopy is ash with alder and birch, with occasional aspen, crab apple, midland hawthorn, wild cherry and wych elm. A dense understorey of hazel, holly and bramble provides perfect habitat for the dormouse, which uses bark from the abundant honeysuckle to construct its nests.

The woodland is awash with butterflies, notably purple hairstreak and silver-washed fritillary. The dark bush cricket can often be heard chirping from sunny patches of bramble. It is the only known breeding site for the barbastelle bat in Wales, which is found alongside seven other bat species. Redstart and wood warbler breed here each summer, joining resident birds including tawny owl, sparrowhawk and buzzard. The site has a rich ground flora, which is especially beautiful in spring, including primrose, wood anemone, golden saxifrage and the diminutive yellow-green flowers of town hall clock.

Pengelli Forest [209].

FFYNONE AND CILGWYN WOODS

CAPEL NEWYDD/NEWCHAPEL, PEMBROKESHIRE — MAP REF 210

Ownership: Private: Calon yn Tyfu **Designations:** SAC, SSSI **Area:** 132ha **Forest type:** mixed	**Forest location:** SN238381 **Explorer Map:** 185 **Ease of access:** Moderate	**Access point:** SN2436138289 ///limped.exist.beep SA37 0HQ

Away from the coast and between the southern tip of the Mynyddoedd Cambrian/Cambrian Mountains and the Mynydd Preseli/Preseli Hills, the landscape of Pembrokeshire and Carmarthenshire has a distinctive beauty. It is one of the least visited and most rural areas in Wales, characterised by pasture fields divided by thick hedgerows, its lanes often brightly lit with yellow laburnum flowers. Woodlands are limited to the few steep-sided valleys whose slopes protected trees from agriculture.

Ffynone and Cilgwyn Woods are one of these precious wooded areas. Owned and managed by a worker cooperative, Calon yn Tyfu (Growing Heart), many of the exotic conifers planted in the 1960s have been felled, although some large Douglas fir remain. A small proportion of ancient woodland remains with veteran oaks, while in places, extensive new areas of 'edible forest' have been created. An amphitheatre with seating has also been constructed within the trees.

The four rivers running through the woods are home to salmon, brown trout and the rare brook lamprey. An attractive waterfall can be visited on the Afon Dulas. A generous car park (Access point) and a network of paths have been installed suiting different abilities. Not to be confused with Cilgwyn Wood [220] in Carmarthenshire.

Chicken of the woods.

A waterfall at Cilgwyn Woods [210].

TY CANOL

FELINDRE FARCHOG, PEMBROKESHIRE MAP REF 211

Ownership: Public: Pembrokeshire Coast National Park Authority and Natural Resources Wales **Designations:** NNR, NP, SAC, SM, SSSI	**Area:** 97ha **Forest type:** temperate rainforest **Forest location:** SN093373 **Explorer Map:** OL35 **Ease of access:** Moderate	**Access point:** SN0927338322 ///exonerate.forecast.hothouse SA41 3XG

Many visitors to the area will head straight to the famous Neolithic chambered dolmen of Pentre Ifan, which dates from 3500 BC. Its 5m-long capstone is impressive, yet just beyond lies the truly wonderful woodland of Ty Canol. The sessile oak, ash and birch trees grow together in a large ancient woodland, prized for its mosses, lichens (more than 400 species) and ferns, which indicate this as a precious fragment of temperate rainforest.

The best access point for the wood is from the north. From a small car park (Access point) walk south to first enter Pentre-Evan Wood, before heading uphill to reach Ty Canol. There, every boulder, tree stem and branch is clad in mosses, lichens and ferns, and all together these create a very special atmosphere. Look for redstart, pied flycatcher and wood warbler in the trees, and sparrowhawk or buzzard overhead.

On a promontory in the south of the wood is an Iron Age enclosure (SN0909136758 | ///happier.taps.renovated), whose ancient defensive walls of boulders are more than 2m high.

ALLT PONTFAEN – COED GELLI-FAWR

PONTFAEN, PEMBROKESHIRE MAP REF 214

Ownership: Public: Natural Resources Wales **Designations:** NP, SAC, SM, SSSI **Area:** 73ha	**Forest type:** temperate rainforest **Forest location:** SN051346 **Explorer Map:** OL35 **Ease of access:** Moderate	**Access point:** SN0583335266 ///shorter.responded.relaxing SA65 9TY

Above the upper reaches of Cwm Gwaun near Llanerch, alder carr woodland grows in the widening valley and its waterlogged peats. An area of this is managed as a nature reserve by the Wildlife Trust of South and West Wales, but access is restricted. However, a good view of the carr woodland and the beautiful valley can be enjoyed from the woods that grow on the steep valley sides to the south of the Afon Gwaun.

The 4km-long finger of sessile oak woodland stretches between Coed Gelli-fawr to the east, continuing westwards through five more woodlands, ending with Allt Pontfaen to the west. The contiguous wooded area is multi-designated for its ancient sessile oak trees, and in particular for a great diversity of lichens. Dormouse inhabits the wood; hunt for the tell-tale signs of hazelnut shells with a hole eaten in their side. Unlike those eaten by wood mouse or vole, the hole will have a smooth inner rim.

From a small lay-by next to a bend in the road (Access point), walk south along the lower reaches of Coed Gelli-fawr. A good path extends along the entire length of the woods. In places, alternative paths extend to the top of the valley sides, such as Coed Tre-Gynon where an Iron Age defensive enclosure (SN0525834528 | ///hers.sprouting.credit) can be seen. Paths are often uneven, and can be steep and muddy in places.

BRECHFA FOREST GARDEN
ABERGORLECH

MAP REF 213

Ownership: Public: Natural Resources Wales **Designations:** **Area:** 1,081ha	**Forest type:** mixed **Forest location:** SN572358 **Explorer Map:** 186 **Ease of access:** Moderate	**Access point:** SN5864033722 ///iron.conga.brimmed SA19 7LX

Nestled near the southern tip of the Mynyddoedd Cambrian/Cambrian Mountains is the unique 'forest garden' at Brechfa. It is one of only three planted in the mid-twentieth century in Britain to test the establishment and growth of non-native tree species. Unlike the forest gardens at Bedgebury (England) and Kilmun (Scotland), Brechfa's origins were less offical, being the personal passion of a local beat forester. Now, it is recognised as a globally significant collection of trees, similar to an arboretum but where the trees are planted in stands rather than as individuals, the aim being to test their productive potential and resilience to environmental change. The humid and sheltered conditions at Brechfa have proven ideal for testing a wide range of species collected from around the world.

The forest garden is laid out in more than 80 plots of different tree species. Alongside giant redwood and coast redwood sampled from California is the internationally endangered Koyama spruce, endemic to Japan but too rare to be of economic value. In contrast, an interesting stand of Japanese red cedar, the species used to construct temples in its native country, grows above the banks of the Afon Gorlech. It has promise as a productive conifer in Britain, although many of the tree stems at Brechfa are swept (curved near their base) mostly likely due to poor genetic selection. Plots throughout the site are clearly labelled with tree species information and are mostly visible from the forest tracks and trail.

The heart of the forest garden is reached after about 3km, following the forest track from the car park. Although a gradual ascent, the track and terrain can be quite uneven. Enjoy a well-earned rest at a picnic table nestling between some magnificent grand fir, offering a great view across the Gorlech valley.

A stand of Japanese red cedar at Brechfa Forest Garden [213].

GALLT Y TLODION

LLANYMDDYFRI/LLANDOVERY, CARMARTHENSHIRE   MAP REF 212

Ownership: Private: Wildlife Trust of South and West Wales **Designations:** **Area:** 17ha	**Forest type:** broadleaved **Forest location:** SN784356 **Explorer Map:** 187 **Ease of access:** Moderate	**Access point:** SN7787334357 ///hydration.guideline.tablets SA20 0RD

A moss-covered bench at Gallt y Tlodion [212].

Gallt y Tlodion [212] is a great site to spot fungi.

Also known as Poor Man's Wood, this woodland was donated to the people of Llanymddyfri/Llandovery by vicar Rhys Prichard (1579–1644), famous for his moral poem 'Cannwyll y Cymry' (The Welshman's Candle). His stipulation was that people could: 'on foot only, enter on the property demised, for the purpose of taking dead wood for fuel, being such amount that they can carry on their backs.'

Now owned by the local wildlife trust, the ancient wood is dominated by sessile oak trees with an understorey of hazel, plus occasional ash, crab apple, holly, rowan and wild service. With the help of a quirky bird hide, look for blackcap, buzzard, lesser spotted woodpecker, nuthatch, pied flycatcher, treecreeper and wood warbler. At the northern boundary is a stream with many interesting bryophytes, including *Hookeria lucens*; a moss that masquerades as a lichen. The main path travels along the foot of the hill, marked by posts with hand-carved characters (although many are now difficult to spot), where ground flora includes bluebell, yellow archangel, wood anemone and wood sorrel. Another climbs the hill soon after the entrance, reaching the upper reaches of the wood where bilberry and moss dominate.

The reserve entrance, which was once marked by an impressive green-woodworked kissing gate (now in disrepair), is accessed by walking from the junction of a minor road with the A40 (Access point), heading past the farm on your left. Shortly after the farm track begins to rise, follow an ancient track, which veers to the left.

The main path through Gallt y Tlodion [212] is almost lost to nature.

TALLEY WOODS

TALYLLYCHAU/TALLEY, CARMARTHENSHIRE

MAP REF 215

Ownership: Public: Natural Resources Wales **Designations:** **Area:** 57ha	**Forest type:** mixed **Forest location:** SN627327 **Explorer Map:** 186 **Ease of access:** Moderate	**Access point:** SN6319732840 ///punters.attend.dots SA19 7BJ

Offering wonderful views across the Talyllychau/Talley valley with its two lakes, a motte and bailey castle, together with the ruins of Talley Abbey, the strenuous routes climbing through these woods are very rewarding. There is a choice of three waymarked routes, all offering glimpses across the valley, and converging at a viewpoint. One of the trails passes through an old arboretum, containing a range of large specimens of both broadleaved and coniferous trees. Some of the conifers above the valley have been clear-felled and are now regenerating as mixed species stands.

The majestic ash near Talley Woods [215].

From the same Access point it is worth exploring the abbey ruins. Founded in the 1180s, this was the first and only abbey constructed by the Premonstratensians in Wales, a monastic order known for their white habits. Their ambitious plans were never fully realised, yet today the abbey's ruined Gothic arches and tower remain very dramatic.

As a further diversion, follow the public footpath beyond the abbey ruins, crossing the valley bottom to see one of the largest ancient ash trees in Wales. It grows next to the path and an old wooden turnstile. The majestic old tree has a whopping stem approaching 6m in diameter (SN6349932907 | ///channel.activates.monitors) and appears to be quite healthy.

The ruins of Talley Abbey and Talley Woods [215].

182 The Forest Guide: Wales

COED CILGELYNNEN

LLANYCHAER, PEMBROKESHIRE — MAP REF 216

Ownership: Private: Woodland Trust
Designations: SSSI
Area: 14ha
Forest type: broadleaved
Forest location: SM982347
Explorer Map: OL35
Ease of access: Moderate
Access point: SM9896034740
///ripe.bands.eyeliner
SA62 5XE

Sessile oak and sycamore, with occasional ash and beech, grow in this thin strip of woodland hidden away from sight on the steep northern slopes above the Cleddau Valley. A footpath starts from the entrance (Access point), where there is space to park a car, and heads upstream through the woods. The valley gradually widens until it reaches a raised bog known as Esgyrn Bottom. A raised bog is formed by centuries of peat deposits and earns the name due to a domed top, which is higher than surrounding land. Raised bogs act like giant sponges and are rich with mosses, grasses and invertebrates. The western end of the wood and the bog is designated an SSSI.

ABER MAWR

MATHRI/MATHRY, PEMBROKESHIRE — MAP REF 217

Ownership: Private: National Trust
Designations: NP
Area: 26ha
Forest type: broadleaved
Forest location: SM884340
Explorer Map: OL35
Ease of access: Easy/Moderate
Access point: SM8846333727
///farms.occupations.jigging
SA62 5HL

Two woodlands line the pretty Afon Mawr as it heads towards the beautiful cove and pebbly beech of Aber Mawr. Pen-yr-allt Wood (west) and Broom Wood (east) are filled with bluebells in spring, and home to many songbirds, while areas of carr woodland support willow and alder in the damp valley bottom. The valley is steeped in history, being a haven for smugglers for many centuries. Fortunately for the wildlife that thrives here, Isambard Kingdom Brunel's plans to create a major railway terminus to connect London with New York never left the drawing board, although some tracks were even laid in preparation during the 1850s. In the 1860s, the next valley to the north at Aber Bach received the first transatlantic telegraph cable and hosted a telegraph station.

A circular walk (3km) can be followed through both woods, allowing a visit to the beach, where sand martins nest in the low cliffs. During very low tides, look for the remains of an ancient petrified forest, though they are less obvious and impressive than at Borth [106]. Most beachgoers will use another car park, which is nearer to the coast, and this provides an alternative access option (SM8839134806 | ///sour.broached.forgiving).

Stone steps climb an embankment at Aber Mawr [217].

COED LETTER

MYDDFAI, CARMARTHENSHIRE — MAP REF 218

Ownership: Private: Woodland Trust Designations: NP Area: 5ha	Forest type: mixed Forest location: SN771305 Explorer Map: OL12 Ease of access: Easy/Moderate	Access point: SN7705130598 ///linen.purifier.heavy SA20 0JD

The nature of this small wood has undergone dramatic changes in recent decades. Once an ancient semi-natural woodland, in 1970 its trees were felled and replaced with Douglas fir and Norway spruce. In 1998, most of these were felled, especially in the south of the wood, and replaced with native broadleaves as part of PAWS restoration (see pp.188–89). More conifers remain at the north of the site near the entrance (Access point) with pockets of naturally regenerating broadleaves. Look for remnants of the former ancient woodland in the ground flora, including enchanter's nightshade, broadleaved helleborine, wood speedwell and yellow archangel. Broad buckler and lady ferns thrive in the damp and shady undergrowth. A permitted path can be followed through the wood, which is fairly level.

NANTYGRONW

CWMDUAD, CARMARTHENSHIRE — MAP REF 219

Ownership: Public: Natural Resources Wales Designations: SM Area: 120ha	Forest type: conifer Forest location: SN361325 Explorer Map: 185 Ease of access: Moderate	Access point: SN3755431234 ///order.trimmer.polygraph SA33 6XA

Forest rides allow the adventurous to explore this quiet coniferous forest, which tracks the pretty Afon Bele along a steep-sided valley upstream of Cwmduad. Among the trees are three earthen round barrows dating from the Bronze Age, although locating them is a challenge without the help of GPS (they are marked 'Tumuli' on OS maps). Access the site near the community centre (Access point).

CILGWYN WOOD

LLANGADOG, CARMARTHENSHIRE — MAP REF 220

Ownership: Public: Natural Resources Wales Designations: SSSI Area: 123ha	Forest type: mixed Forest location: SN750298 Explorer Map: OL12 Ease of access: Moderate	Access point: SN7443729748 ///like.ironic.riverboat SA19 9LF

Lying just outside the border of the Bannau Brycheiniog/Brecon Beacons National Park, this charming woodland possesses a sense of calm and tranquillity. A single 3km-long waymarked circular trail, which includes some moderate climbs, passes through a mix of large Douglas fir, Norway spruce and western red cedar, with stands of mature beech trees. The wood is often attractively coloured, whether by bluebells in spring or by beech leaves in autumn. A small area of the site is designated as an SSSI for its unusual geology.

Not to be confused with Ffynone and Cilgwyn Woods [210] in Pembrokeshire.

Continuous cover forestry at Glynaeron Forest [221].

GLYNAERON FOREST

ROSEBUSH, PEMBROKESHIRE

MAP REF 221

Ownership: Private
Designations: NP
Area: 58ha
Forest type: conifer

Forest location: SN107298
Explorer Map: OL35
Ease of access: Moderate

Access point: SN1066429513
///grandson.galloped.removals
SA66 7QW

Created as a commercial plantation in the late 1950s, mostly with Sitka spruce and some Japanese larch, Glynaeron is one of two plantations in the local area belonging to the same landowner. In contrast to other plantations in the Preseli Hills, both of these award-winning forests are now managed according to the principles of continuous cover forestry. Their transformation from stands of single species and age to those of mixed species and diverse ages is slow and gradual, as forest managers aim to maximise tree growth for timber while beginning to underplant with diverse productive conifer and broadleaved species. This will also improve habitat for wildlife and lead to greater economic and environmental resilience. The owner has sponsored long-term bird monitoring and ringing, revealing the presence of both pied and spotted flycatcher, marsh tit, firecrest and lesser redpoll. Goshawk breed among the conifers, while peregrine falcon is often seen overhead. Permissive access to the forest is granted by the owner (i.e. there is no public right of way), which can be enjoyed by following the network of rides that climb through the forest. The views across the hills are expansive. Pantmaenog Forest [222] lies to the west.

PANTMAENOG FOREST

ROSEBUSH, PEMBROKESHIRE MAP REF 222

Ownership: Private
Designations: NP, SM
Area: 259ha
Forest type: conifer

Forest location: SN085311
Explorer Map: OL35
Ease of access: Moderate

Access point: SN0703829636
///regress.barman.able
SA66 7QY

Supported by the Pembrokeshire National Park Authority, the private owner of this large forest provides extensive public access to visitors. There are more than 12km of forest rides to follow, with dedicated routes for cyclists and horse riders. The commercial forest grows prominently on the southern slopes of the Mynydd Preseli/Preseli Hills, extending from a disused quarry near a large car park, and up the valley of the Afon Syfni to reach 490m above sea level at its upper ridge. Look for red kite and buzzard overhead, and goldcrest and firecrest in the canopy of the Sitka spruce. Given the distances involved, and altitude, dress for hillwalking.

COED PENCASTELL

BLAENWAUN, CARMARTHENSHIRE MAP REF 223

Ownership: Private
Designations: SM
Area: 105ha
Forest type: broadleaved

Forest location: SN257290
Explorer Map: 185
Ease of access: Moderate

Access point: SN2489028185
///meanest.heap.showering
SA33 6DS

Like nearby Ffynone and Cilgwyn Woods [210], Coed Pencastell exists as a thin wooded dingle hidden within a deep valley of the otherwise-treeless agricultural landscape. Ash and hazel line the very steep sides of the Afon Asen and nearby Afon Cynin, and accessing either on foot is a challenge. One option is to walk to Castell Pencastell, an Iron Age defensive hill fort occupying the triangular-shaped promontory overlooking the Afon Asen. Its 1.5m-deep ramparts are hidden among the trees. Beyond here, the footpath descends to the valley floor. The terrain is often overgrown, steep, muddy, and difficult to follow.

COED CWM TAWEL

CYNWYL ELFED, CARMARTHENSHIRE MAP REF 224

Ownership: Private
Designations: SM
Area: 42ha
Forest type: broadleaved

Forest location: SN382255
Explorer Map: 185
Ease of access: Moderate

Access point: SN3896025880
///pulsing.doghouse.disputes
SA33 6AR

A single track climbs the woods above the meanders of the Afon Gwili and the echoing whistle of the locomotive reaching the end of the Gwili heritage railway at Danycoed Halt (meaning 'below the woods'). The wood is dominated by sessile oak, with ash and alder on deeper and wetter soils. Listen for the distinctive calls of wood warbler, which sounds like a coin spun on a glass tabletop, its rapid rising trills coming to a sudden stop. Just inside the entrance to the wood are some Second World War anti-invasion defences, including a pillbox, thought to have been constructed to defend water supplies.

VELINDRE WOOD

LLYS-Y-FRÂN, PEMBROKESHIRE   MAP REF 225

Ownership: Public: Natural Resources Wales **Designations:** **Area:** 19ha	**Forest type:** conifer **Forest location:** SN043266 **Explorer Map:** OL35 **Ease of access:** Moderate	**Access point:** SN0404724454 ///times.taped.mixing SA63 4RR

Velindre Wood is only one of several named woods surrounding Llys-y-frân reservoir in the foothills of the Preseli Hills. The reservoir can be circumnavigated (11km) on foot, while owners Welsh Water offer a range of water- and land-based activities to cater for all tastes from a recently renovated visitor centre. Velindre Wood lies at the northern tip of the water body. Many of its trees were larch but have been felled due to phytophthora disease, and it now mainly comprises some large Douglas fir, Sitka spruce and oak trees. The wood provides important habitat for roosting bats, which can be seen feeding over the water (designated SAC and SSSI) on summer evenings.

ST DAVIDS

TYDDEWI/ST DAVIDS, PEMBROKESHIRE   MAP REF 226

Ownership: Various **Designations:** NP, SM **Area:** 12ha **Forest type:** broadleaved	**Forest location:** SM748253 **Explorer Map:** OL35 **Ease of access:** Easy	**Access point:** SM7489325384 ///bordering.flukes.term SA62 6PX

A green vein of trees flows through the heart of Tyddewi/St Davids, Britain's smallest city, their canopy covering 13 per cent of its urban area. The scattered trees, copses, spinneys and dells line the steep valley sides of the Afon Alun/River Alyn and surround the city's green spaces and buildings. One of these is the enormous medieval ruin of St Davids Bishop's Palace, built to help cater for the thousands of pilgrims who flocked here when a twelfth-century pope decreed that two visits to Tyddewi/St Davids was worth one pilgrimage to Rome. Next door, beyond a babbling stream, lies the cathedral, built on the site of a sixth-century monastery. Inside, look under the flip-up seats of the choir stalls for a beautiful wooden-carved green man with oak sprigs emerging from his mouth.

The ruins of St Davids Bishop's Palace.

Conifer trees undergoing removal at Coed Maenarthur [124].

Plantations on Ancient Woodland Sites

When Wales was subjected to afforestation in the twentieth century with a view to creating a strategic timber reserve, in addition to planting at huge scale on the treeless uplands, many existing woodland sites at lower altitudes thought to be of poor economic value were felled to be replaced with productive conifers. Many of these sites were ancient woodlands or semi-natural woodlands, and we now recognise the considerable value that they once held for biodiversity, and also for landscape and society (e.g. archaeology). Such sites are known as Plantations on Ancient Woodland Sites (PAWS).

Definition

A PAWS site is a forest site believed to have been continuously wooded for over 400 years, and currently with a canopy cover of more than 50 per cent non-native conifer tree species.

Following the felling, planting with conifers and timber-focussed management, it might be expected that all ecological and cultural value has been lost forever, yet with sensitive management

some PAWS sites can respond well to restoration. Remnants of ancient semi-woodland can sometimes be seen in a few surviving veteran trees crowded in by conifers, standing deadwood, or the presence of ancient woodland indicator plants such as bluebells, yellow archangel and wood spurge. Such remnants provide a shining light and hope for the future, as restoration is more likely to be successful, even if this may take many decades to come to reality.

Restoration needs to be undertaken sensitively and patiently by the forest owner. Conifer trees can be removed around surviving broadleaves, called 'halo thinning', and cleared away from streams and archaeological features.

Among sites included in this guide, some 11 are PAWS sites. One of the largest PAWS sites undergoing restoration is a 50ha area of Long Wood [133]. Another is a large block of 31ha in Gwydir Forest [34], near to Artist's Wood [33]. There are 19 ancient woodland sites that are presently undergoing restoration. Coed Penglanowen [116] and neighbouring Old Warren Hill [117] are both examples of sites that have been restored. Another and the largest restored site featured, covering 106ha, is at Wyndcliff Wood [315].

CASTLE WOODS

LLANDEILO, CARMARTHENSHIRE

MAP REF 227

Ownership: Private: Wildlife Trust of South and West Wales Designations: IPA, NNR, SM, SSSI Area: 19ha	Forest type: broadleaved Forest location: SN615218 Explorer Map: 186 Ease of access: Easy/Moderate	Access point: SN6275922051 ///officer.fishnet.words SA19 6BN

Castle Woods comprises two blocks of woodland on the Dinefwr Estate. While Newton House and much of the parkland (which hosts many ancient trees) is owned by the National Trust, most of the woodland to the south, overlooking the Afon Tywi/River Towy, is owned by the local wildlife trust. The largest block includes the remains of Castell Dinefwr; the seat of generations of Princes of Wales, but which fell to ruin by 1523. To the east, in a smaller woodland block nearer to town, the trees surround the diminutive church at Llandyfeisant, assigned to the abbey at Talyllychau/Talley (see Talley Woods [215]) in the fourteenth century.

The woods grow on steeply sloping ground, carpeted with bluebell, dog's mercury and wood anemone. Also in spring, the mysterious pink-blushed flowers of toothwort emerge, growing as a parasite on the roots of hazel and wych elm. The main canopy is pedunculate oak, accompanied by ash, beech and sycamore. Look for silver-washed fritillary and speckled wood butterflies. The woodland supports a rich diversity of bryophytes, including tree lungwort. Birdsong fills the air and the ancient trees support breeding populations of nuthatch, redstart, treecreeper, tawny owl, both species of flycatcher and all three species of woodpecker. In winter, bring binoculars to spot large numbers of wildfowl on the river floodplain below the woods.

Park on the side of a small road next to Llandeilo Bridge (Access point) or otherwise in a car park next to the fire station. Visitors wanting to explore the nearby deer park and grounds owned by the National Trust should use (SN6256322564 | ///electric.snug.croutons).

Castell Dinefwr and Castle Woods [227].

PORTH CLAIS

PORTHCLAIS, PEMBROKESHIRE — MAP REF 228

Ownership: Private: National Trust **Designations:** IPA, NP, SM, SSSI **Area:** 1ha **Forest type:** broadleaved	**Forest location:** SM739243 **Explorer Map:** OL35 **Ease of access:** Easy	**Access point:** SM7398524302 ///messy.even.elsewhere SA62 6RR

Adjacent to the pretty Roman-era harbour and the ancient holy well of Pistyll Dewi, a tiny yet dense woodland of stunted oak, sycamore and hawthorn nestles from the sea winds on the far-most tip of Tyddewi/St Davids Peninsula. As the most westerly woodland in Wales, the site quite regularly features as a refuge for rare migrant birds, blown off course and relieved to find sanctuary.

LOUGHOR

LLANDYFAN, CARMARTHENSHIRE — MAP REF 229

Ownership: Private **Designations:** NP **Area:** 1ha **Forest type:** broadleaved	**Forest location:** SN668178 **Explorer Map:** 178 **Ease of access:** Moderate	**Access point:** SN6715817738 ///people.skate.reminds SA19 6UA

A ring of trees surrounds the emergence of Nant Llygad Llwchwr at the cave of Llygad Llwchwr ('Eye of the Loughor'). Its waters have travelled 6.4km from a natural underground reservoir below Y Mynydd Du/Black Mountain. The cave system is famous among caving enthusiasts, both novices under supervision and experienced cave divers who explore its deep, flooded caverns.

The cave and river exit (SN6687017800 | ///currently.lure.tabs) are reached after a short walk from a small lay-by (Access point), crossing a field with large sinkholes either side. Many of the ash trees surrounding the sinkholes are suffering terribly from dieback. Continue and the sound of rushing water soon provides a hint of the destination. Access to the cave entrance is barred by the landowner due to its use as a source for a commercial bottled water product, although a stile is provided to allow visitors to see the river emerging from underground. Listen for cuckoo and look for dipper along the tree-lined river.

Llygad Llwchwr or the Eye of the Loughor.

The Great Glasshouse and the arboretum of the National Botanic Garden of Wales [230].

NATIONAL BOTANIC GARDEN OF WALES

PORTHYRHYD, CARMARTHENSHIRE　　　　　　　　　　　MAP REF 230

Ownership: Private: National Botanic Garden of Wales **Designations:** HPG **Area:** 6ha	**Forest type:** mixed **Forest location:** SN523179 **Explorer Map:** 178 and 186 **Ease of access:** Easy	**Access point:** SN5185617875 ///making.partly.frail SA32 8HG

The National Botanic Garden of Wales aims to be an internationally renowned centre for biodiversity conservation and research. Most of the trees in its arboretum have been planted since it was created in 2000, and although still young, they are thriving. Together with collections from Wales, those from China, South America and North America can be explored. The arboretum lies next to Waun Las National Nature Reserve, a mosaic of beautiful wildflower-rich hay meadows and pastures. Excellent interpretative signage helps showcase the trees and other plants, which can be accessed using the all-ability paths. Entry charge (assistant dogs only).

BATHESLAND WOOD

ROCH, PEMBROKESHIRE MAP REF 231

Ownership: Private: National Trust
Designations: NP
Area: 8ha
Forest type: broadleaved

Forest location: SM859209
Explorer Map: OL36
Ease of access: Easy/Moderate

Access point: SM8578620120
///mental.vintages.same
SA62 6BD

Bathesland Wood is on the Southwood Estate, gifted to the National Trust in 2003. It forms part of a diverse landscape including farmland and one of the longest stretches of sand in Wales at Newgale Beach. Follow a trail from the car park (Access point), keeping an eye on the hedgerows for yellowhammer, whose breeding populations are nationally declining. The wood contains many beautiful ash trees clad in mosses and ferns. The trail can be continued to complete a 4km circuit, which returns to the start via the beach.

COED WERN DDU

LLANLLWCH, CARMARTHENSHIRE MAP REF 232

Ownership: Private: Wildlife Trust of South and West Wales
Designations:
Area: 1ha

Forest type: broadleaved
Forest location: SN372179
Explorer Map: 178
Ease of access: Easy/Moderate

Access point: SN3737217936
///cycles.firelight.offers
SA33 5BU

Coed Wern Ddu runs as a bright green slither for 300m along either side of a pretty brook. Its sessile oak trees with downy birch and wych elm, plus some occasional large beech trees, have a vibrant understorey of guelder rose, gorse, holly and rowan. This is a shady wood, rich with bryophytes and ferns. Look for the rare Turner's threadwort, one of our most attractive liverworts. Early purple orchid is often seen in spring, while foxglove thrives in midsummer.

WITHYBUSH WOODS

HWLFFORDD/HAVERFORDWEST, PEMBROKESHIRE MAP REF 233

Ownership: Public: Pembrokeshire County Council
Designations:
Area: 7ha

Forest type: broadleaved
Forest location: SM966186
Explorer Map: OL36
Ease of access: Easy

Access point: SM9626718897
///iterative.reprints.lakes
SA62 4BN

Adjacent to an industrial trading estate on the northern outskirts of Hwlffordd/Haverfordwest, this small broadleaved woodland is filled with birdsong during spring and summer. 'Withybush' is another name for the willows that thrive here. A 1.5km-long accessible footpath, including sections with guide rails for the visually impaired, allows visitors to enjoy the woodland, and a large pond and wetland area. Look for hart's-tongue fern and early purple orchid growing in the shade of the sycamore trees.

Hart's-tongue fern.

CARMEL

PENTRE-GWENLAIS, CARMARTHENSHIRE — MAP REF 234

Ownership: Public: Natural Resources Wales **Designations:** NNR, SAC, SSSI **Area:** 35ha	**Forest type:** broadleaved **Forest location:** SN601164 **Explorer Map:** 178 **Ease of access:** Moderate/Difficult	**Access point:** SN6052616431 ///lied.drumbeat.harmlessly SA18 3JQ

The limestone bedrock, which has been quarried on this site since the Middle Ages, supports a rich and diverse flora. At its heart is a large old quarry and a lake. Fed solely by underground water, which comes and goes with the seasons, this type of lake is known as a turlough and is the only one of its kind in Britain. There are several old lime kilns across the site.

Ash, which is terribly sickened by dieback, dominates the patchy woodland cover, although sycamore is thriving. The trees are dispersed by smaller quarries, spoil heaps and wildflower grassland. Bluebell, dog's mercury, herb-Paris, lily-of-the-valley, toothwort, ramsons and wood anemone carpet the woodland floor, with hart's-tongue fern thriving in shady areas.

A single waymarked trail can be followed around the site. This is quite strenuous in places, with loose rock and some steep climbs. The view from the top of the quarry sides are very fine, but beware the precipitous sides. The Wildlife Trust of South and West Wales cares for the western end of the NNR near Carmel village.

The quarry sides at Carmel [234] loom out of the mist.

Bluebells at Green Castle Woods [235].

GREEN CASTLE WOODS

LLANGAIN, CARMARTHENSHIRE — MAP REF 235

Ownership: Private: Woodland Trust **Designations:** **Area:** 50ha	**Forest type:** broadleaved **Forest location:** SN391167 **Explorer Map:** 178 **Ease of access:** Easy/Moderate	**Access point:** SN3922816421 ///represent.podcast.warns SA33 5BG

Unlike the dramatic ruins of Castell Llansteffan [237] at the mouth of the Afon Tywi/River Towy, 8km further upstream only subtle signs remain of the grand motte and bailey of Green Castle, which once stood here overlooking the river. Meanwhile, the remains of a fortified seventeenth-century manor house can still be seen among the trees (SN3952016710 | ///nimbly.replenish.abode).

From the car park (Access point), three different entrances to the woods can be used, from where a range of waymarked trails can be followed. Near the car park, wood pasture and new woodland planting can be accessed via paths suitable for all abilities. The older woods to the north can be visited as part of a circular route, although this is steep in places and often muddy. Some very pretty sections pass through ancient sunken lanes and along old hedgerows, creating a palpable sense of walking back in time.

The oak, ash and downy birch trees, with alder thriving in wet areas, host a rich diversity of woodland birds. Birdsong can be cacophonous, especially at dawn, while displays of bluebell, primrose and wood anemone in spring are glorious.

MINWEAR FOREST

ARBERTH/NARBERTH, PEMBROKESHIRE　　　　　MAP REF 236

Ownership: Public: Natural Resources Wales **Designations:** NP, SM **Area:** 144ha	**Forest type:** mixed **Forest location:** SN054138 **Explorer Map:** OL36 **Ease of access:** Easy/Moderate	**Access point:** SN0543913810 ///incorrect.detonated.worth SA67 8BL

Lining the south bank of the Cleddau estuary at the top of its tidal range, Minwear Forest was once well placed to provide charcoal and timber for local industries. Its oak trees were used for shipbuilding, and the charcoal made from coppices sent to a nearby iron foundry. During the twentieth century, many of the broadleaves were felled and replaced with conifers.

A waymarked trail passes through the northern part of the wood, including a viewpoint with benches overlooking the estuary, easily reached from a large lay-by (Access point). A car park at the north-eastern end (SN0595114290 | ///certified.trades.hubcaps) lies closer to a well-preserved medieval castle ringwork (SN0618113481 | ///observe.sprinkle.fearfully), which can be discovered above a steep cwm of the Penglyn Brook (but note, no waymarking). The long-distance (96km) and circular Landsker Borderlands Trail passes through the forest.

CASTELL LLANSTEFFAN

LLANSTEFFAN, CARMARTHENSHIRE　　　　　MAP REF 237

Ownership: Private **Designations:** SM **Area:** 4ha **Forest type:** broadleaved	**Forest location:** SN352100 **Explorer Map:** 177 **Ease of access:** Moderate	**Access point:** SN3549010853 ///buzzer.polygraph.foods SA33 5LW

Commanding dramatic views across the Tywi estuary and surrounding countryside, Castell Llansteffan was an important medieval stronghold, which passed hands several times between English and Welsh defenders during the twelfth and thirteenth centuries. A small broadleaved wood grows on the very steep slopes south of the substantial ruins. General public access to the wood is not permitted (unlike the castle grounds) but the Wales Coast Path passes through its lower reaches. Look for cormorants drying their wings on the tree branches, and heron fishing on the shore. Walk from the public car park (Access point) at the far end of Llansteffan beach.

A stonechat perches on a young alder tree near the shoreline in front of Castell Llansteffan [237].

LITTLE MILFORD WOOD

MADDOX MOOR, PEMBROKESHIRE

MAP REF 238

Ownership: Private: National Trust	**Forest location:** SM964117	**Access point:** SM9677411586
Designations:	**Explorer Map:** OL36	///mull.paramedic.echo
Area: 23ha	**Ease of access:** Easy/Moderate/	SA62 4LQ
Forest type: mixed	Difficult	

The woodland at Little Milford, growing above the estuary, salt marsh and mudflats of the West Cleddau estuary, has existed since the eleventh century, if not earlier. Remains of limekilns and an old coal mine can be discovered under the canopy, while the sessile oak trees have for centuries been regularly harvested and coppiced for local industries. The more accessible areas were replaced with conifers during the twentieth century, and although many of these trees have been felled there are areas remaining, with western red cedar and other conifers growing alongside ash, sycamore, willow and hazel.

Little Milford Wood is contiguous with Hook Wood to the east, designated an SSSI and owned by Natural Resources Wales. Its sessile oak trees continue to cloak the steep slopes above the Cleddau estuary and can be explored by following a public footpath along the shoreline past a deep curve in the river known as Hook Bight.

The views across the estuary are splendid. Cormorant, heron and little egret are commonly seen near the water's edge, while the wood supports pied flycatcher, redstart and treecreeper.

West Cleddau estuary and Little Milford Wood [238].

BLAENANT Y GWYDDYL

GLYN-NEDD/GLYNNEATH, NEATH PORT TALBOT — MAP REF 239

Ownership: Private: Wildlife Trust of South and West Wales **Designations:** **Area:** 10ha	**Forest type:** temperate rainforest **Forest location:** SN885076 **Explorer Map:** OL12 **Ease of access:** Difficult	**Access point:** SN8834406851 ///makeovers.finest.exact SA11 5BD

Blaenant y Gwyddyl is among a cluster of temperate rainforest sites in the upper Afon Neath valley, including two sites across the county border into Powys (Central-East) found along its tributaries Nedd Fechan and Afon Mellte (Coed Nedd [200] and Coed Mellte [199]).

Sessile oak and birch trees dominate the canopy of this ancient wood, with ash, small-leaved lime, sycamore and wych elm. Hazel and hawthorn form an understorey, which has been heavily grazed in the past, with bracken dominating in summer, while alder thrives in wet flushes, which are generally richer in plants. Look for pied flycatcher, redstart and wood warbler.

Two waterfalls can be discovered towards the north end of the wood, the upper one falling from a small cliff face (SN8854407859 | ///deposits.topical.

Hazel and waterfall.

succumbs). Paths are indistinct and can be overgrown in summer, and the terrain is steep. Park along Lon y Nant road (Access point), and continue on foot up Glynmelyn Road past the farm to reach the wood.

COLBY WOODLAND GARDEN

AMROTH, PEMBROKESHIRE — MAP REF 240

Ownership: Private: National Trust **Designations:** NP **Area:** 64ha **Forest type:** broadleaved	**Forest location:** SN158080 **Explorer Map:** OL36 **Ease of access:** Moderate	**Access point:** SN1581008091 ///enhancement.bagels.witty SA67 8LU

The Colby Estate is perched on the end of Pembrokeshire's coal deposit. Like other mines that once operated nearby, it employed children to haul the coal out of its narrow seams when it was mined in the late eighteenth century. The industrial heritage of the site is clear to see, with remains camouflaged among the attractive woodland garden, whose construction first began in 1870.

Visitors are attracted to the walled garden, a tea room and a glorious flowering wet meadow. Three trails can be followed around the estate, all of which demand some quite steep ascents. Larger areas of woodland can be explored north of the road. In the woodland east of the Newt Pond (SN1566007869 | ///sampled.rejoins.deferring) is one of the tallest Japanese red cedar (also known as sugi) trees in Britain, towering 39m. Otter, bats and a whole host of woodland birds including dipper and grey wagtail thrive in the woodland and its streams. Parking is charged but there is free entrance to the site.

WESTFIELD PILL

NEYLAND, PEMBROKESHIRE — MAP REF 241

Ownership: Private: Wildlife Trust of South and West Wales **Designations:** **Area:** 7ha	**Forest type:** broadleaved **Forest location:** SM963067 **Explorer Map:** OL36 **Ease of access:** Easy/Moderate	**Access point:** SM9608407341 ///hike.marine.chilling SA73 1JF

This waterside nature reserve is a little gem. More than 150 species of bird have been recorded, including firecrest and hoopoe in the oak, ash and hazel trees, plus little grebe, night heron, osprey and water rail on the estuary or at the small lake and its reedbeds. More than 30 species of butterfly have been spotted too. Adder and grass snake enjoy sunny glades, while at night four species of bat feed on flying invertebrates. Otter is often seen early in the day. Among the woodland edge, look for the rare wild service tree.

A small lay-by provides convenient access from the north (Access point), while it is also possible to access the site from the south, parking at the end of the road near the Neyland yacht marina (SM9679805795 | ///given.jungle.fattening). A Sustrans route known as the 'Brunel Trail' passes right through the reserve, following Isambard Kingdom Brunel's original Great Western Railway line.

CWM-DU GLEN AND GLANRHYD PLANTATION

PONTARDAWE, NEATH PORT TALBOT — MAP REF 242

Ownership: Public: Neath Port Talbot Council **Designations:** LNR **Area:** 15ha	**Forest type:** broadleaved **Forest location:** SN715045 **Explorer Map:** 165 **Ease of access:** Moderate	**Access point:** SN7202904032 ///mixing.appendix.prune SA8 4AD

Attractive broadleaved woods follow the Upper Clydach River and the gorge of Cwm-Du Glen, extending north for 1km from the heart of Pontardawe and culminating in a dramatic waterfall. Oak and ash lean over the tumbling water, with ferns and wood anemone (plus invasive rhododendron) growing along its banks. Look for dipper and kingfisher by the river, and flocks of tits in the trees. The former grounds of Glanrhyd House lie above the gorge and can be included as part of a wonderful circular route. The house (demolished in 1968) was built by Arthur Gilbertson in 1878 and during the First World War was used as an auxiliary Red Cross hospital to help with the convalescence of wounded soldiers. A large specimen of coast redwood is one of several trees dating from the creation of the original estate. Park in the town and access the site at Herbert Square, near the main crossroads.

Dipper.

Arboreta

Living collections of trees, or arboreta (singular 'arboretum'), are scattered across Wales, although there are perhaps fewer examples than might be expected, given the excellent growing conditions for trees in the country. There were once many more arboreta, often established by enthusiastic Victorian landowners when rare exotic plants were the height of fashion. Old arboreta can still be discovered, like at Coed Aberartro [60] and Talley Woods [215], though don't expect any interpretation at these sites.

Incense cedar is native to the north-west USA.

Bodnant Garden [12]
An arboretum and pinetum that houses several champion trees, notably a giant redwood, which reaches more than 50m tall. Other large trees include an Atlas cedar, Nordmann and Greek firs, and bishop and Mexican white pines.

Goat Field
A small arboretum of native trees, which lies next to Coed Deri [144].

Monkey puzzle.

Gregynog [148]
Large specimens of cedar, copper beech, giant redwood and monkey puzzle can be enjoyed on this private estate managed by the Gregynog Trust.

Brechfa Forest Garden [213]
Technically not an arboretum, as it contains stands of trees rather than individual specimens. The forest garden provides foresters with more information about how exotic trees interact with each other under forest conditions.

Dyffryn Gardens (not featured)
An Edwardian garden managed by the National Trust, which includes 9ha collection of exotic trees (ST0961772873 | ///voter.party.mountains).

National Botanic Garden of Wales [230]
A young arboretum, yet already Wales' premier collection of trees, and a dedicated visitor attraction for arboreal enthusiasts. Along with native Welsh provenances (origins) of trees are sections dedicated to China, South America and North America.

Naylor Pinetum [146]

A collection of more than 100 conifer species, with some very large coast redwoods. Open only to members of the Royal Forestry Society or during organised events.

Parc Bute (Bute Park)
(not featured)

Behind Cardiff Castle in the centre of the city, this arboretum has more than 40 champion trees (notable for their size) and a huge diversity of broadleaved species. Park in North Road (ST1766377412 | ///modest.humid.format).

Plas Newydd [20]

Part of a larger estate that offers stunning views of Eryri/Snowdonia, this arboretum features many Australasian species.

Swansea Botanical Garden
(not featured)

Set within the grounds of Singleton Park, this botanic garden has some attractive specimen trees, though the focus is more on the herbaceous and tropical plants growing in its glasshouses (SS6345191970 | ///wing.horns.comical).

Treborth Botanic Garden
(not featured)

Owned by Bangor University, this arboretum has recently benefited from a revival in management. Among the fine Douglas firs, yews and various spruce species are some unusual species including coffin tree, Hungarian oak and sawtooth oak. A collection of Welsh provenances of native trees is underway (SH5506271050 | ///lifeboats.inversion.thrashing | LL57 2RQ).

Coast redwood at Brechfa Forest Garden [213].

WEST WILLIAMSTON

WEST WILLIAMSTON, PEMBROKESHIRE MAP REF 243

Ownership: Private: Wildlife Trust of South and West Wales
Designations: NP, SSSI
Area: 4ha

Forest type: broadleaved
Forest location: SN028058
Explorer Map: OL36
Ease of access: Moderate

Access point: SN0323405880
///quicker.sage.processor
SA68 0TL

Huge numbers of waders and wildfowl congregate on a promontory at the confluence of the Afon Caeriw/River Carew and Afon Cresswell. Inland from the creeks, tidal mudflats and salt marsh, a small woodland of oak, sycamore and ash, with patchy blackthorn scrub, grows among the old quarries and spoil heaps. It is ideal habitat for the brown hairstreak; one of Britain's rarest butterflies. Adults of the elusive butterfly congregate high up in the canopy of prominent trees (often ash) where they mate, and the females then descend to the blackthorn to lay their eggs. The eggs hatch in spring to coincide with the leaf burst of the blackthorn, the larvae (caterpillars) feeding on the tender young foliage.

A small car park is found along a lane at the west end of the village, just before a cattle grid. From there, walk along the private road, which doubles as a public footpath, to reach the nature reserve. Binoculars are recommended, both to spot the tell-tale spiralling flight of courting brown hairstreak butterflies and to appreciate the wonderful diversity of birds.

LLYN FACH

HIRWAUN, NEATH PORT TALBOT MAP REF 244

Ownership: Private: Wildlife Trust of South and West Wales
Designations: SSSI
Area: 17ha

Forest type: mixed
Forest location: SN906036
Explorer Map: 166
Ease of access: Difficult

Access point: SN9269503102
///firelight.juror.goat
CF44 9UE

Hidden among the Sitka spruce of the large plantation of Pen y Cymoedd [245], Llyn Fach is one of two small nutrient-poor lakes below the impressive crags of Craig y Llyn. The north-facing cliffs were created by glacial processes during the last Ice Age. On the periphery of the reserve, areas of larch felled due to disease are providing opportunities for natural regeneration, while remaining mature conifer trees harbour crossbill and goldcrest. Nightjar and cuckoo make good use of the open wooded habitat. The most interesting tree area is found below the cliffs and scree, coloured crimson by scattered rowan trees in autumn and perhaps a hidden rare whitebeam or two. Wilson's filmy fern and beech fern grow in damp, shady hollows.

From the lay-by on the A4061 (Access point) head downhill to begin a walk across moorland and using forest tracks to reach the nature reserve (6km return). Visitors should be competent navigators. A good view of the site can be enjoyed from the top of the crags, following a section of the 58km-long Coed Morgannwg Way.

Rowan berries.

PEN Y CYMOEDD

HIRWAUN, NEATH PORT TALBOT MAP REF 245

Ownership: Public: Natural Resources Wales Designations: SM Area: 1,011ha	Forest type: conifer Forest location: SN894033 Explorer Map: 166 Ease of access: Moderate	Access point: SN9218002512 ///wolf.unspoiled.reduction CF44 9UF

Swathes of larch trees have been felled at Pen y Cymoedd in recent years due to the outbreak of *Phytophthora ramorum*; a type of water mould that kills the trees. Nonetheless, Sitka spruce and lodgepole pine dominate this large commercial forest planted up to 600m altitude on the Rhigos mountain and surrounding territory. The nature reserve of Llyn Fach [244] is a hidden jewel, several prehistoric remains exist, while the 76 turbines of Pen y Cymoedd, Britain's highest-altitude windfarm, are unmissable. Park in a small lay-by next to the entrance to the windfarm (Access point) and follow Sustrans Route 47 into the trees. Route 887 (linking with Afan Forest Park [254]) joins to the south-west, offering opportunities to extend a walk or cycle.

Young female cones of Sitka spruce.

CWM CLYDACH

CLYDACH, SWANSEA MAP REF 246

Ownership: Private: RSPB Designations: Area: 50ha Forest type: broadleaved	Forest location: SN677033 Explorer Map: 165 Ease of access: Easy/Moderate	Access point: SN6834302581 ///screen.amplified.presenter SA6 5TA

The oak, ash and hazel woods gracing the steep valley of the Afon Clydach are rich with wildlife. Coal mining was once a major industry in the area but the trees in the cwm were protected thanks to its precipitous terrain. The fast-flowing river regularly erodes its banks, creating new habitats for pioneer species. Wood sorrel is common on the woodland floor in spring, and a good diversity of fungi appear in autumn, including chantarelle. Look for pied flycatcher, redstart and treecreeper among the main woodland, and bullfinch, marsh tit, lesser redpoll and siskin among the purple-tinted alder trees near the river, where grey wagtail and dipper feed. Silver-washed fritillary and speckled wood butterflies grace sunlit glades.

A 5km trail is waymarked to the top of the cwm and back. Initially easy going on level paths near the valley bottom, it becomes more strenuous further up the cwm. For the adventurous, the route of the 16km circular Cwm Clydach Walk can be followed, which passes through the reserve and beyond, taking in more woodland further upstream before returning via open ground along the hilltops to the east.

CWM MELIN-CWRT

MELIN-CWRT/MELINCOURT, NEATH PORT TALBOT MAP REF 247

Ownership: Private: Wildlife Trust of South and West Wales
Designations: SM
Area: 5ha
Forest type: temperate rainforest
Forest location: SN824017
Explorer Map: 166
Ease of access: Moderate
Access point: SN8213301991
///plausible.lied.builder
SA11 4BE

Visitors have come to this small wooded cwm at Melin-cwrt/Melincourt to see the spectacular 24m-high waterfalls for centuries, including J. M. W. Turner, who produced a half-coloured sketch of the falls in 1795. While the famous artist was at work, just downstream from the falls one of the last charcoal-fuelled ironworks in Britain was still active (it closed in 1808). Its remains can be seen today, including a blast furnace, wheel pit and leat. Today, the ancient upland oak woodland is managed as a nature reserve by the local wildlife trust. The wood is a fabulous site for ferns, which grow near the brook and under the spreading canopies of the sessile oak, small-leaved lime and silver birch trees, with 20 species recorded, including brittle bladder-fern, green spleenwort, hay-scenter buckler-fern and Wilson's filmy fern. The site is also rich with mosses, including flagellate feathermoss, and a distinctive leafy liverwort, western pouncewort. Bluebells carpet the floor in spring, with hosts of other flowering plants appearing during the summer, including enchanter's nightshade. Look for redstart, pied flycatcher and wood warbler.

PEMBREY FOREST

PEN-BRE/PEMBREY, CARMARTHENSHIRE MAP REF 251

Ownership: Public: Natural Resources Wales
Designations:
Area: 981ha
Forest type: conifer
Forest location: SN388025
Explorer Map: 164 and 178
Ease of access: Easy/Moderate
Access point: SN4013100127
///alas.mainly.exam
SA16 0EJ

Situated between the waters of Carmarthen Bay to the west, a Ministry of Defence firing range to the north and a local airport to the east, Pen-bre/Pembrey Forest is a large area of conifers growing along the Carmarthenshire coastline. Its Corsican pine trees, among the few tree species that can thrive in such poor soils, were planted in the first half of the twentieth century to help stabilise the sand dunes and to produce timber. The forest grows partly on the site of an old explosives factory, where TNT was produced between the First and Second World Wars.

In sunlit glades, look for common lizard, grizzled skipper and marsh fritillary butterflies. In dune slacks (damp depressions between dune ridges), southern marsh and pyramidal orchids grow, and even the rare fen orchid. Common crossbills take full advantage of the huge quantity of pine seeds, while goldcrests are common.

The Access point is a large car park (charge), which is part of nearby Pen-bre/Pembrey Country Park. A 7km-long trail can be followed, which makes use of the old trackbeds of munitions trains that once ran up and down the coastline, passing blast tunnels and bunkers, now home to greater horseshoe bat. The Wales Coast Path passes through the site.

Pen-y-Bedd Wood, a 32ha outlying block of Corsican pine, grows next to the main access road (SN4195701377 | ///quite.majors.client). Offering a picnic site and a short trail, it is home to the Pembrey Conservation Trust, whose volunteers help manage wildlife across Pembrey Forest.

CRAIG GWLADUS

ABERDULAIS, NEATH PORT TALBOT — MAP REF 248

Ownership: Public: Neath Port Talbot Council Designations: Area: 55ha	Forest type: broadleaved Forest location: SS765995 Explorer Map: 165 Ease of access: Easy/Moderate	Access point: SS7684499770 ///acted.aside.track SA10 8LG

Previously the site of a popular zoo, and with a longer history of coal mining, Craig Gwladus is now a country park and nature reserve beloved by local people. The beech, oak, silver birch and sycamore trees grow on a steep south-east-facing slope among curious rock formations. Many of the larger trees host 'fairy doors' installed by children. Several paths can be followed, some of which involve climbing and steps. Enjoy expansive views across the Cwm Nedd/Neath Valley.

PENLLERGARE FOREST

PENLLERGARE, SWANSEA — MAP REF 249

Ownership: Public: Natural Resources Wales Designations: SM Area: 193ha	Forest type: mixed Forest location: SN624004 Explorer Map: 165 Ease of access: Moderate	Access point: SS6250299567 ///class.proper.issue SA4 9HL

Many of the trees in the ancient woodland that once covered the hills north of Penllergare were felled and replaced with conifers during the last century. These are now being removed in an effort to restore ancient woodland habitat, while managers also deal with the increasing impact of ash dieback, and encroaching invasive cherry laurel and rhododendron. A pleasant 5km route can be followed through the forest. A prehistoric defensive earthwork is found in the north-east of the site overlooking the valley of Nant y Crimp. On the southern and opposite side of the M4 motorway lies Penllergare Valley Woods [252].

HOLYLAND WOOD

PENFRO/PEMBROKE, PEMBROKESHIRE — MAP REF 250

Ownership: Private Designations: Area: 4ha Forest type: broadleaved	Forest location: SM996016 Explorer Map: OL36 Ease of access: Easy/Moderate	Access point: SM9965601612 ///gracing.marsh.mammals SA72 4SR

This small privately owned wood is managed by the Pembroke 21C Community Association and is much loved by local people. Among the oak and ash trees, a fairy walk has spontaneously appeared. Areas of alder and willow carr woodland grow in wet areas, while small-leaved lime and wych elm thrive on drier ground. Look for marsh tit and willow warbler in the carr woodland. A good diversity of ferns, including broad buckler, lady, male and soft shield, thrive in the damp and shady conditions. Immediately adjacent to the west of the site is Pembroke Upper Mill Pond; an LNR managed by the local wildlife trust, although it is not accessible to the public.

PENLLERGARE VALLEY WOODS

PENLLERGAER, SWANSEA — MAP REF 252

Ownership: Private: The Penllergare Trust **Designations:** HPG **Area:** 110ha	**Forest type:** mixed **Forest location:** SS627987 **Explorer Map:** 165 **Ease of access:** Easy/Moderate	**Access point:** SS6235199221 ///oven.gangs.healers SA4 9GY

Penllergare Valley Woods is an important historical landscape designed by botanist and pioneering photographer, John Dillwyn Llewelyn (1810–1882). Parts of the original estate have been at threat from development, but a significant area is now cared for by the Penllergare Trust, whose staff and volunteers are busy restoring the landscape to its former glory, often with the help of Llewelyn's daguerreotype images. Unlike neighbouring Penllergare Forest [249], the mixed woods on the estate were never replanted for commercial timber production.

An attractive waterfall below the upper lake, where Daubenton's bats feed in the evening, has recently been restored. Keep an eye out for otter and water vole in Lower Lake, and kingfisher along the Afon Llan. The woods brighten with spring flowers, including bluebell and wood anemone, and host summer visiting birds, including blackcap and whitethroat.

A number of trails and paths can be followed, the longest (3km) of which runs the length of the woods and involves some sections with steps. Paths can be muddy in places. The site lies very close to Junction 47 of the M4.

GNOLL ESTATE COUNTRY PARK

CASTELL-NEDD/NEATH, NEATH PORT TALBOT — MAP REF 253

Ownership: Public: Neath Port Talbot Council **Designations:** HPG **Area:** 20ha	**Forest type:** mixed **Forest location:** SS768975 **Explorer Map:** 165 **Ease of access:** Easy/Moderate	**Access point:** SS7664797619 ///those.papers.ozone SA11 3DT

This country park is a popular picnic site with local people and can be busy at certain times thanks to its wide range of attractions. Enjoy an easy stroll around the tree-lined perimeter of the Fishpond and its cascades, or to escape the crowds walk north-east through Mosshouse Wood. An easy-to-follow path climbs gently from the visitor centre, rising more steeply as it approaches Mosshouse Reservoir, built to provide water for the town. The route once passed through a larch plantation, but the conifers have been felled due to an outbreak of *Phytophthora ramorum* and broadleaves are now being encouraged to regenerate.

To the south-east, Brynau Wood (72ha) can also be reached from the same access point. The former ancient woodland is being restored by The Woodland Trust, and extended as part of an ambitious natural flood management scheme. Parking charge.

KILVEY HILL

ABERTAWE/SWANSEA, SWANSEA   MAP REF 255

Ownership: Public: Swansea Council Designations: SM Area: 62ha	Forest type: mixed Forest location: SS671945 Explorer Map: 165 Ease of access: Easy/Moderate	Access point: SS6641094678 ///narrow.burst.bared SA1 7FR

Lying on the eastern fringes of Abertawe/Swansea city, the wooded heights of Kilvey Hill are much loved as a green space by local people. Its management is supported by a community group of active volunteers. Part of the site has been under threat from the development of a major leisure facility.

Blackcap, chiffchaff, whitethroat and willow warbler join the chorus of resident songbirds during summer months. Peregrine can be watched hunting overhead, while nightjar can be seen in clearings at dusk. One of the oldest and largest trees on the site is a Monterey pine, but most of the site is covered in Scots pine trees, which are being slowly cleared to favour a wide range of broadleaves.

Reaching the top of the hill is quite strenuous but involves easy-to-follow tracks. From the summit, enjoy dramatic views across the city, the docks and Bae Abertawe/Swansea Bay. Near the Access point are the remains of a copper works established in 1736, while near the top of Kilvey Hill is a Romano-British enclosure dating from AD 74–AD 410.

Looking over Abertawe/Swansea city from Kilvey Hill [255].

AFAN FOREST PARK

PONTRHYDYFEN, NEATH PORT TALBOT — MAP REF 254

Ownership: Public: Natural Resources Wales **Designations:** SM **Area:** 781ha	**Forest type:** conifer **Forest location:** SS813942 **Explorer Map:** 165 and 166 **Ease of access:** Easy/Moderate	**Access point:** SS8208395121 ///clubbing.rooftop.insist SA13 3HG

Afan Forest Park is a large forest area created in the 1970s and now celebrated as one of the top destinations for thrill-seeking mountain bikers, while also offering a selection of walking trails. Those along the valley bottom following the river and old railway lines near the visitor centre are typically easy. These include Sustrans Route 887, which travels past on its 30km route from Port Talbot into the heart of Pen y Cymoedd [245] forest. Another is a section of Saint Illtyd's Walk, a 103km long-distance footpath. Many walkers enjoy routes from the visitor centre north of Cwm Afan, such as along the ridge of Gyfylchi, which offers spectacular views to the sea, and Bae Abertawe/Swansea Bay, although the terrain is steep and can be quite demanding. The forest covers many small prehistoric sites, including cairns, enclosures and a standing stone. Look out for goshawk, while adders bask along the edges of forest rides.

The ex-colliery site and visitor centre are included within the Valleys Regional Park – see also Cwmcarn Forest [320] (South-East), and Dare Valley Park [268] and Parc Slip [277] (South-Central) – and also forms part of the National Forest for Wales. The Access point is the visitor centre (parking charge), which includes a cafe, toilets and bike shop.

CRYMLYN BOG AND PANT Y SAIS

PORT TENNANT, SWANSEA — MAP REF 256

Ownership: Public: Natural Resources Wales **Designations:** LNR, NNR, SSSI **Area:** 33ha	**Forest type:** broadleaved **Forest location:** SS715943 **Explorer Map:** 165 **Ease of access:** Easy/Moderate	**Access point:** SS6856794206 ///crops.racks.hang SA1 7BG

Crymlyn Bog and Pant y Sais are two sites that make up a single spectacular NNR and SSSI, the largest low-lying wetland in Wales. Pockets and fringes of wet woodland add valuable structure to the reed and sedge beds. Willow species dominate, while alder, birch and gorse are common. Alder buckthorn, foodplant of the brimstone butterfly, thrives here too. Reed bunting, stonechat and willow warbler take advantage of the trees, and in spring the calls of cuckoo echo across the trees and reeds. In shady areas, look for the elegant and large fronds of the royal fern. A visitor centre for Crymlyn Bog can be found on Dinam Road (Access point), offering trails through wet woodland with sections of boardwalk.

A separate Access point allows limited access to Pant y Sais (SS7126194062 | ///acid.payout.birdcage). A circular accessible boardwalk (500m) can be followed, and halfway round it is possible to depart and walk along the towpath of the Tennant canal. The nearest parking for Pant y Sais is along the residential street Heol Yr Ysgol/School Road.

STACKPOLE ESTATE

BOSHERSTON, PEMBROKESHIRE

MAP REF 257

Ownership: Private: National Trust **Designations:** HPG, IPA, NNR, NP, SAC, SM, SPA, SSSI **Area:** 186ha	**Forest type:** mixed **Forest location:** SR977959 **Explorer Map:** OL36 **Ease of access:** Easy/Moderate	**Access point:** SR9797996511 ///musician.youth.outs SA71 5DA

The natural diversity of the Stackpole Estate is stunning, ranging from limestone sea cliffs and dunes to freshwater lakes, parkland and woodland. Watch for chough near the coast, while the woodland supports one of Britain's strongest populations of the endangered greater horseshoe bat. Many of the trees are less than 200 years old, planted by the owners of Stackpole Court, the Cawdors. While the land was planted with conifers during the 1960s, broadleaves are now being encouraged instead, with sycamore thriving alongside oak and ash. Many old beech and sweet chestnut trees are scattered across the estate.

There are several locations to start enjoying this stunning 1,200ha estate. The Access point given is at Stackpole Court (charge), as this provides easy access to the woodland of Lodge Park, part of which falls within the NNR, and a nearby deer park.

Nearby, another car park (SR9796696518 | ///folds.harvest.forwarded) provides access to Castle Dock Woods. Its mixed conifers and broadleaves can be enjoyed via a circular trail. This site is popular with mountain bikers, and is steep in places.

The car park at Bosherston (SR9668094850 | ///brink.official.chuck) lies nearer to the coast and allows access to nearby North Hill Wood, which also falls within the multi-designated conservation area. A circular waymarked trail provides a wonderful walk through the woods and around the shores of the lily ponds, where otter and heron can be seen. The sands of Broad Haven lie at the furthest point, offering a welcome diversion.

Above: Grey heron.

Right: The lily ponds at the Stackpole Estate [257].

WHITEFORD BURROWS

LLANMADOC, SWANSEA   MAP REF 258

Ownership: Private: National Trust **Designations:** IPA, NL, NNR, SAC, SPA, SSSI **Area:** 32ha	**Forest type:** conifer **Forest location:** SS439942 **Explorer Map:** 164 **Ease of access:** Moderate	**Access point:** SS4397693502 ///bigger.wires.strutted SA3 1DJ

The stunning 3km-long stretch of Whiteford Sands is one of the Gower's most unspoilt beaches. Further inland lie the extensive salt marshes of the Afon Llwchwr/River Loughor. Sandwiched between are the sand dunes of Whiteford Burrows, multi-designated for their rare flora, including early marsh orchid and dune gentian, among 250 other species. Several large and discrete stands of Corsican pine extend along their length, connected by the Wales Coast Path. Climb Cwm Ivy Tor to enjoy extensive views across the trees and dunes as far as Berges Island. The pines were planted for their timber and are being progressively removed to help restore dune habitat, although some are being retained for landscape reasons. A wide range of birds can be seen among the tree canopies, while binoculars will allow distant views of waders and waterfowl. Private car park (charge). Note there are no local facilities.

PARC LE BREOS CWM AND COED Y PARC

PARKMILL, SWANSEA   MAP REF 259

Ownership: Public: Natural Resources Wales **Designations:** IPA, NL, SAC, SM, SSSI	**Area:** 149ha **Forest type:** mixed **Forest location:** SS536900 **Explorer Map:** 164	**Ease of access:** Easy/Moderate **Access point:** SS5383489605 ///arriving.widget.processes SA3 2EH

The valley known as Parc le Breos Cwm possesses a very rich history, dating back to prehistoric times. Cathole Cave, created from a nature fissure in the limestone rocks of the gorge, has revealed the bones of mammoth and woolly rhinoceros, plus giant deer from the Neolithic, and human tools from the Bronze Age. The nearby burial chamber of Parc Cwm long cairn (a cromlech) was discovered by accident in the nineteenth century, and found to contain the remains of at least 40 people. In more modern times, the land was a medieval deer park managed by the lords of Gower. Later still, in the mid-nineteenth century, lime was produced from the rocks to improve soil fertility and for use in mortar for buildings, and the remains of a lime kiln and quarry can still be seen.

Coed y Parc/Park Wood falls under the Gower Ash Woods SAC (like nearby Coed Nicholaston [262] and Oxwich Wood [263]). It is dominated by ash, with oak and sycamore also present. Large areas were planted with conifers in the nineteenth and twentieth centuries, although many of these are now being thinned. The wood is rich with wildlife, and especially noted for populations of

Looking out from a lime kiln at Parc le Breos Cwm [259].

Oak trees and a lime kiln at Parc le Breos Cwm [259].

lesser and greater horseshoe bat. The site can become busy at certain times of the year, so visit early in the day to enjoy the tranquillity and glorious birdsong.

A car park (Access point) provides good access to the valley, a surfaced path allowing an easy route up through the gorge, passing very close by to the cromlech. The woods cling to the steep valley sides, so exploring them and Cathole Cave requires more effort (the entrance to the cave is now gated to protect roosting bats). At the head of the valley, three different paths can be followed through the wood, two of which include stretches of the long-distance (56km) Gower Way.

MILL WOOD

PENRICE, SWANSEA

MAP REF 260

Ownership: Public: Natural Resources Wales Designations: HPG, NL Area: 164ha	Forest type: broadleaved Forest location: SS488881 Explorer Map: 164 Ease of access: Easy/Moderate	Access point: SS4930088200 ///unrated.coach.universes SA3 1LN

The wood takes its name from a sixteenth-century corn mill that was once part of the Penrice Estate. Many of its ancient ash and oak trees were replaced by the Forestry Commission in the 1950s, planting Norway spruce, silver fir, western hemlock and beech. While these are being gradually removed, signs of the original eighteenth-century landscape can also be seen, including an avenue of lime trees and a gloom of yews, resulting in a woodland with plenty of character. The ruins of the water mill, including its grindstone, can be seen, along with a fishpond. Just outside the wood's boundary, a medieval castle ringwork is found to the south, and the Norman-period stone-built Penrice Castle to the north. Look for wild daffodil, bluebell and ramsons, and male fern in shady spots. The wood falls within the Gower NL.

COED NICHOLASTON

PENRICE, SWANSEA MAP REF 262

Ownership: Public: Natural Resources Wales Designations: IPA, NL, SAC, SSSI Area: 38ha	Forest type: broadleaved Forest location: SS513879 Explorer Map: 164 Ease of access: Moderate	Access point: SS5026488121 ///reason.demotion.folk SA3 2HN

Two woodlands grow at either end of the spectacular Oxwich National Nature Reserve. To the south of one of the most popular beaches on the Gower Peninsula is the privately owned Oxwich Wood [263], while Coed Nicholaston lies to the north. Both form part of the Gower Ash Woods SAC but fall outside the NNR.

Coed Nicholaston is dominated by ash, with pedunculate oak, small-leaved lime and sycamore also present. The understorey includes spindle and butcher's broom, with many notably large wild clematis. The lime-rich soils support dense carpets of bluebell and ramsons in spring, and rarer plants including herb-Paris and purple

Right: The giant limbs and deadwood of this ancient oak tree at Coed Nicholaston [262] provides habitat for countless other species.

Below: Ramsons.

gromwell. Both lesser and greater horseshoe bats thrive in the woods.

A 4km circular walk starts in a small parking area (Access point), with room for only two cars. The path starts with some stone steps and is level at first but becomes more uneven and hilly. Occasional glimpses through the trees across Oxwich Bay are tremendous, while a small viewing area at the east end of the wood is kept free of trees, offering extensive views of the marsh, burrows, Oxwich Wood [263] and the open sea. Mill Wood [260] is close by.

PENNARD CLIFF AND NORTHILL WOODS

PARKMILL, SWANSEA — MAP REF 261

Ownership: Private: National Trust **Designations:** IPA, NL, SAC, SM, SSSI **Area:** 28ha	**Forest type:** broadleaved **Forest location:** SS544885 **Explorer Map:** 164 **Ease of access:** Easy/Moderate	**Access point:** SS5446389198 ///zooms.sizes.essay SA3 2EH

On a spur overlooking the languorous meanders of Pennard Pill and beyond to one of the Gower's most beautiful beaches at Three Cliffs Bay, stand the dramatic ruins of Pennard Castle. The Norman defensive position was originally constructed with timber before being replaced with impressive stone walls complete with arrow slits. Two ash-dominated woods with oak and sycamore line the valley sides, one on the steep limestone cliffs below the castle, the other (Northill) on the west side of the river. Watch for heron and sand martin, and feeding bats at dusk. Paths can be followed on either side along the lush valley, while it is a short, steep climb through the trees to reach the castle ruins. A circular walk can be followed, crossing the river at stepping stones (water level permitting) used by the Wales Coast Path. There are several possible access points, but the easiest is from a large car park to the north of the site in Parkmill (charge).

Pennard Castle.

Oxwich Wood [263] seen beyond the dunes and expansive sands of Oxwich Bay.

OXWICH WOOD

OXWICH, SWANSEA

MAP REF 263

Ownership: Private	**Forest type:** broadleaved	**Access point:** SS4993086438
Designations: IPA, NL, SAC, SM, SSSI	**Forest location:** SS504859	///unhelpful.books.vertical
	Explorer Map: 164	SA3 1LU
Area: 41ha	**Ease of access:** Moderate	

This site is one of two woods found either side of the marshes and reedbeds of Oxwich National Nature Reserve, the other being Coed Nicholaston [262] to the north. Both fall within the Gower Ash Woods SAC, designated because of their rich plant life and associated species. Oxwich Wood grows on steep north-east facing slopes, its trees of ash, oak and sycamore shaped by sea winds. Minor tree components include wild service and wych elm, and the rare rock whitebeam. Butcher's broom occurs in the shrub layer, its prickly leaves giving rise to its traditional use for sweeping chopping blocks. Another of its names, knee holly, is obvious when its red fruits appear in late summer.

At the wood's northern end lies the remains of Oxwich Castle; not actually a castle but a grand Tudor fortified manor house. Among the trees below are an old quarry and kiln (SM).

There are two main entrances to the wood. From the large car park (charge) used by beachgoers, follow the Wales Coast Path southwards past a hotel (whose proprietors own the wood) and enter the wood, which immediately climbs very steeply. Alternatively, starting from a small lay-by (Access point) on the hill below Oxwich Castle reduces some of the climb, though the footpath can be overgrown in summer. This path joins the coast path after 800m and can be followed to the far tip of the wood at Oxwich Point. The views across Oxwich Bay are breathtaking.

This is the smallest of the regions, incorporating the counties of Merthyr Tydfil, Cardiff, Vale of Glamorgan, Bridgend and Rhondda Cynon Taf, yet proportionally it has the second greatest amount of forest cover at almost 16 per cent, which is above average for Wales. It has been deeply scarred by the exploitation of coal at an industrial scale, but is now the focus for regeneration, in terms of both the social and natural kind (see Green to black and back, pp.230–31). Former coal sites include Clydach Vale [273], Cwm Saerbren [270], Dare Valley Park [268], Parc Slip [277]

and Parc Coedwigaeth Penpych [269]. In contrast, the northern tip of the region falls within the Bannau Brycheiniog/Brecon Beacons National Park, while there are plenty of wildlife-rich sites to explore elsewhere, such as the heronries at Coed Llwyn [281]. Coastal woodlands include the pretty Porthkerry Woods [285] and Cwm Colhuw [286], the most southerly site in this guide, while inland, the spectacular waterfalls at Parc Coedwigaeth Penpych [269] are worth a visit.

A waterfall at Penpych [269].

SOUTH-CENTRAL

MERTHYR TYDFIL, CARDIFF, VALE OF GLAMORGAN, BRIDGEND, RHONDDA CYNON TAF

SITES 264–286

COED TAF FAWR

MERTHYR TUDFUL/MERTHYR TYDFIL, RHONDDA CYNON TAF

MAP REF 264

Ownership: Public: Natural Resources Wales
Designations: NP
Area: 448ha

Forest type: conifer
Forest location: SO002131
Explorer Map: OL12
Ease of access: Easy/Moderate

Access point: SO0025113131
///frown.glares.momentous
CF48 2HY

Coed Taf Fawr is a large forest in Cwm Taf, planted for timber production and to protect and clean drinking water in three reservoirs constructed to supply Cardiff during the early twentieth century. The forest is well served by Garwnant Visitor Centre (Access point), including cafe and toilets, which is the start to several waymarked trails catering for all abilities. Sculptures line some of the trails, which also pass by the ruins of Wern Farm, while those climbing higher into the forest offer attractive views. A 5km tree-lined circuit of Llwyn-on Reservoir, the largest of the three, can also be followed. The Taff Trail, linking Cardiff with Aberhonddu/Brecon, also passes through the woodland. The site falls within Fforest Fawr UNESCO Global Geopark, which is bringing investment to the area.

DARREN FACH

CEFN-COED-Y-CYMMER, MERTHYR TYDFIL

MAP REF 265

Ownership: Private: Wildlife Trust of South and West Wales
Designations: IPA, NP, SSSI
Area: 7ha

Forest type: broadleaved
Forest location: SO018106
Explorer Map: OL12
Ease of access: Difficult

Access point: SO0239808972
///fend.impact.filer
CF48 2HP

On the east side of the A470, north of Merthyr Tudful/Merthyr Tydfil, is the impressive crag known as Darren Fawr. The weathered limestone cliffs are popular with rock climbers, who are challenged by multiple climbs along the crag's five main walls.

At the northern end of the crags lies a small woodland within an area designated as an SSSI for its rich biodiversity. Wales' rarest tree, Ley's whitebeam, grows at Darren Fach. There are believed to be only 17 known specimens in the wild, while it is possible to see a cultivated specimen at the National Botanic Garden of Wales [230]. The species is believed to be a hybrid of two species of *Sorbus*, one of the grey whitebeams and the more-common rowan, and was discovered by nineteenth-century clergyman Reverend Augustin Ley.

Reaching the wood is demanding. Do not attempt to access the woodland directly from the A470, as the short track does not cross open-access land. Instead, park in a small lay-by (Access point) 2km to the south, and climb the zigzag path past Danydarren Quarry to reach a wall that extends along the top of the crags. Walk carefully, avoiding the precipitous cliff edges, aiming for a track on the far side of the hill that descends into the wood, only becoming obvious just above it (SO0209210838 | ///coconuts.wooden.yours). Warning: the route is exposed with rough terrain and requires competency in navigation with map and compass.

CWM TAF FECHAN

MERTHYR TUDFUL/MERTHYR TYDFIL, MERTHYR TYDFIL

MAP REF 266

Ownership: Private: Wildlife Trust of South and West Wales **Designations:** HPG, IPA, LNR, SSSI **Area:** 12ha	**Forest type:** broadleaved **Forest location:** SO041091 **Explorer Map:** OL12 **Ease of access:** Moderate	**Access point:** SO0408607337 ///like.enjoyable.hunter CF47 8RE

Stretching between Cefn Coed and Pontsarn bridges over the Taf Fechan, this beautiful ancient woodland lies surprisingly close to Merthyr Tudful/Merthyr Tydfil. It is designated as an LNR and SSSI, particularly for its assemblages of bryophytes. Woodland birds, including buzzard, great spotted woodpecker, jay, nuthatch, tawny owl and treecreeper, live and breed among the ash, beech and birch trees. Alder and grey willow grow near the tumbling water, where dipper, kingfisher and grey wagtail thrive. Look for speckled wood and small pearl-bordered fritillary butterflies.

The most satisfying route to enjoy the wood is to follow an 8km trail from Castell Cyfarthfa/Cyfarthfa Castle, where there is ample and free parking, completing a circuit that takes advantage of paths on both banks of the Taf Fechan. The thirteenth-century castle of Morlais provides an interesting diversion at the top end of the route. The Taff Trail traces the upper and western fringes of the wood, offering an easier route with a view over the wooded valley.

DARE VALLEY PARK

CWMDARE, RHONDDA CYNON TAF

MAP REF 268

Ownership: Public: Rhondda Cynon Taff County Borough Council and Natural Resources Wales **Designations:**	**Area:** 15ha **Forest type:** broadleaved **Forest location:** SN966020 **Explorer Map:** 166 **Ease of access:** Easy/Moderate	**Access point:** SN9851302598 ///advantage.frown.grudging CF44 7RG

Prior to 1853, the valley of the Afon Dâr/River Dare was a remote farming community in Cwm Cynon, but the discovery of coal transformed the area (see Green to black and back, pp.230–31). Several pits were amalgamated over time to create the Bwllfa Colliery, which remained open until the 1990s. Dare Valley Park includes remnants of mining heritage, although the site has been levelled, including pit winding gear and coal trams. Follow the blue trail through oak woodland to reach the head of the cwm, where there is a peregrine viewing platform with excellent views of the crags of Tarren y Bwllfa. An easier route is waymarked, which keeps to the valley floor. A short distance to the south-east, the Dare Valley Community Woodland offers brilliant engagement with local people, including adventure camps for children and a forest school. The site is included within the Valleys Regional Park (see also: Afan Forest Park [254], Cwmcarn Forest [320] and Parc Slip [277]).

A pair of peregrine falcons.

GETHIN WOODLAND PARK

ABERCANAID, MERTHYR TYDFIL — MAP REF 267

Ownership: Public: Natural Resources Wales
Designations: SM
Area: 856ha
Forest type: mixed
Forest location: SO044032
Explorer Map: 166
Ease of access: Easy/Moderate
Access point: SO0505603383
///edit.shiny.spins
CF48 1YZ

Once the site of Gethin Colliery, the valley and wooded hillsides above now provide enjoyment and recreation for visitors, space for nature, and timber production. Norway spruce grows productively here, while natural regeneration with diverse species is favoured in many areas. The park is particularly popular with mountain bikers, whose needs are served by a huge range of trails and even a shuttle service for downhill racers. A range of paths and trails can be walked, whether climbing westwards to the top of the hill where prehistoric cairns can be found, or northwards towards the old pit, railway and an ironworks. Near the car park (charge), a woodland pond, which once provided water to Gethin Pit No2, is now managed as a small nature reserve.

CWM SAERBREN

TREHERBERT, RHONDDA CYNON TAF — MAP REF 270

Ownership: Public: Natural Resources Wales
Designations: SSSI
Area: 41ha
Forest type: mixed
Forest location: SS932978
Explorer Map: 166
Ease of access: Moderate/Difficult
Access point: SS9385698177
///geese.chill.howler
CF42 5HT

Overlooking Rhondda Fawr, Cwm Saerbren is a sheltered corrie rich with wildlife, backed by dramatic sandstone cliffs rising to 510m above sea level. It forms part of the Mynydd Ty-isaf SSSI. Among the crags look for a frondage of ferns, including beech, brittle bladder, broad buckler, lady and mountain. Keep an eye out above for peregrine falcon. To reach the site, park in the street near the level crossing (Access point) next to the railway station, before walking across the line and following a choice of informal tracks and paths into the steep-sided cwm.

GOITRE COED FACH

QUAKER'S YARD, MERTHYR TYDFIL — MAP REF 271

Ownership: Private: Woodland Trust
Designations: SM
Area: 4ha
Forest type: broadleaved
Forest location: ST092967
Explorer Map: 166
Ease of access: Easy/Moderate
Access point: ST0954996534
///sourcing.paces.clap
CF46 5AT

Nestling inside a bend of the Afon Taf/River Taff, Goitre Coed Fach is a small part of a longer stretch of riverside woodland. The old trackbed of the Merthyr Tramroad runs alongside, once serving coal and iron industries, but now a popular walking and cycling route (the Taff Trail). Beech trees line the river, while oak, ash and birch grow above. Park in any of the residential streets near the Access point. A well-trodden path climbs into the wood, sections of which can be steep and precipitous, or otherwise saunter along the tramroad. The bridge over the river at the west end of the wood is a Scheduled Monument.

PARC COEDWIGAETH PEN-PYCH

BLAENRHONDDA, RHONDDA CYNON TAF — MAP REF 269

Ownership: Public: Natural Resources Wales
Designations:
Area: 1,011ha
Forest type: conifer
Forest location: SN909009
Explorer Map: 166
Ease of access: Moderate
Access point: SS9235999097
///swipes.weaned.duet
CF42 5DS

Rhondda's very own table mountain of Pen-pych lies at the head of Rhondda Fawr, surrounded by conifer plantations. A viewpoint that looks up to the spectacular waterfall in Cwm Lluest is a popular destination, only a short walk (1.2km) from a large car park (Access point). The falls remain invisible until the final approach, although the spectacular falls in neighbouring Nant yr Ychen draw the eye for much of the climb. A longer 9km circular walk can be followed past the observation point, climbing above the waterfall to reach the mountain top after a short scramble, although a recent landslip has made this perilous and the path may remain shut for some time. The longer route continues through the trees to reach the head of the valley, where there are more waterfalls to behold before descending its eastern flanks, the return route offering good views of the old colliery, a disused railway tunnel and the forest.

The waterfall at Cwm Lluest appears through winter low cloud at Parc Coedwigaeth Pen-Pych [269].

ST GWYNNO FOREST

TYLORSTOWN, RHONDDA CYNON TAF   MAP REF 272

Ownership: Public: Natural Resources Wales
Designations: SM
Area: 1,157ha

Forest type: conifer
Forest location: ST032958
Explorer Map: 166
Ease of access: Moderate

Access point: ST0325295845
///scrapping.dressy.sectors
CF39 0RA

St Gwynno is a large forest stretching along the north side of Rhondda Fach. A range of informal routes can be followed through the conifers, with several interesting features to explore, including Clydach Reservoir, a waterfall, and a Roman marching camp at Twyn y Bridallt (ST0017398235 | ///hunk.limitless.forget). The prominent and partially tree-clad cultural icon of Tylorstown Tip (ST0196995598 | ///spots.owner.trees) lies nearby, the huge coal tip offering an interesting diversion, with dramatic views across Rhondda Fach.

CLYDACH VALE

TONYPANDY, RHONDDA CYNON TAF   MAP REF 273

Ownership: Public: Rhondda Cynon Taff County Borough Council and Natural Resources Wales
Designations:

Area: 579ha
Forest type: mixed
Forest location: SS969931
Explorer Map: 166
Ease of access: Easy/Moderate

Access point: SS9835792771
///depravity.bleat.talents
CF40 2XX

In place of the old colliery shafts and spoil heaps of the old Cambrian Colliery (one of the largest in the Rhondda valley), Cwmclydach Country Park (owned by the local borough and managed by the local community) has transformed the valley into a green oasis. Planting of broadleaved trees, natural regeneration and landscaping work around two lakes ('Top Lake' and 'Bottom Lake') can be enjoyed via accessible paths, in an area covering 13ha. A range of trails can be followed from the Bottom Lake (Access point) next to a cafe. Those in the valley bottom are easy to follow, including a circuit that reaches Top Lake and a memorial to miners, including the disaster of 1965 when 31 men lost their lives due to an underground explosion. Other more-demanding trails can be followed, which explore the valley sides, reaching the extensive Sitka spruce plantations of Mynydd Ton and Mynydd Bwllfa (Natural Resources Wales), offering stimulating views across the vale. Look for southern marsh and bee orchids in open areas, and listen for the distinctive call of cuckoo, which often echoes across the valley each spring.

Hazel catkins and rain.

SPIRIT OF LLYNFI

MAESTEG, BRIDGEND   MAP REF 274

Ownership: Public: Natural Resources Wales
Designations:
Area: 29ha

Forest type: broadleaved
Forest location: SS854920
Explorer Map: 166
Ease of access: Easy/Moderate

Access point: SS8512291725
///goat.curvy.gaps
CF34 0AN

Some 60,000 broadleaved trees have been planted on this community woodland site overlooking the Llynfi Valley, included within the National Forest for Wales. Once the site of Coegnant Colliery and Maesteg Washery, the soils of the area are heavily polluted, yet the trees are flourishing and the emerging woodland much loved by local people. A striking 2.7m-tall oak sculpture of a miner stands at the site's centre, which is dissected by forest roads, paths and running trails, sections of which are steep further up the hill. Park in any nearby residential street and follow Sustrans Route 885 into the trees.

ALLT Y RHIW

BLACKMILL, BRIDGEND   MAP REF 275

Ownership: Public: Coity Wallia Commoners Association
Designations: SAC, SSSI
Area: 71ha

Forest type: broadleaved
Forest location: SS930856
Explorer Map: 151 and 166
Ease of access: Moderate

Access point: SS9327986310
///amuses.beginning.rail
CF35 6DP

Ancient sessile oak woods line both banks of the Afon Ogwr/River Ogmore downstream of Blackmill. Collectively known as Blackmill Woodlands, both are subject to high-status conservation designations. This type of woodland is rare this far south in Wales. Accordingly, it is much drier than true temperate rainforest sites further west and north.

The woodlands have been managed as common land since the Middle Ages, with different associations managing blocks on either side of the river. Their gnarly forms are a result of centuries of lopping and coppicing for firewood, under ancient permissions known as estover rights. The larger block to the south, known as Allt y Rhiw, is dissected by several paths and an old railway. The easiest access is from a small lay-by

Fern croziers or fiddleheads.

(Access point) where the Nant Cwm-dwr exits the wood. The site is steep throughout and paths can become overgrown in summer.

The block on the north side of the valley is known as Craig Tal y Fan (managed by the Llangeinor Commoners Association) and can be accessed from town. Sustrans Route 4 runs parallel to this wood's lower edge.

FFOREST FAWR

TONGWYNLAIS, CARDIFF

MAP REF 276

Ownership: Public: Natural Resources Wales
Designations: SM, SSSI
Area: 105ha
Forest type: broadleaved
Forest location: ST133832
Explorer Map: 151
Ease of access: Easy/Moderate
Access point: ST1417883864
///enter.lowest.defend
CF15 7JR

Fforest Fawr ('great forest' in English) is no longer as extensive as its name suggests, but lying close to the northern periphery of Cardiff City, it provides wonderful refuge to both people and wildlife. Mainly broadleaved and dominated by beech, the forest also has small pockets of conifers, especially western hemlock, and lights up with bluebells, ramsons and wood anemones each spring. About 18ha of this site is designated an ancient semi-natural woodland.

The iron industry that came to dominate this landscape for centuries, fuelling the Industrial Revolution, first began locally in 1560. Coal, haematite (an iron-rich mineral) and limestone were all mined here. Evidence of shallow mining pits and spoil heaps can be seen everywhere, while two deeper mines dug to follow rich veins of haematite await discovery along waymarked trails. Picked out by hand, with a little help from dynamite, the Three Bears Cave and Blue Pool are worth a visit (but do not be tempted to enter!). The Cambrian Way also passes through the trees, while a short sculpture trail can be enjoyed near the main car park (Access point).

At the south-west tip of the forest, commanding extensive views across the Taff Gorge, stands Castell Coch, also known as *castrum rubeum* 'the red castle'. Built on top of the remains of a medieval castle, the Gothic-style castle dates from the Victorian era. Its prominent conical towers are surrounded by steep-sided beech woods designated as an SSSI. It is possible to explore the castle and grounds (managed by Cadw) but charges apply (alternative access: ST1326882688 | ///stage.dogs.jazzy).

The Three Bears Cave at Fforest Fawr [276].

224 The Forest Guide: Wales

PARC SLIP

TONDU, BRIDGEND

MAP REF 277

Ownership: Private: Wildlife Trust of South and West Wales
Designations:
Area: 61ha
Forest type: broadleaved
Forest location: SS885839
Explorer Map: 151
Ease of access: Easy/Moderate/Difficult
Access point: SS8815184176
///minivans.fear.vision
CF32 0EW

Now a nature reserve, Parc Slip was once an opencast coal mine. The valley has been greened with the addition of lakes (with bird hides), reedbeds, flower-rich grassland and young woodland. Lots of waymarked paths criss-cross the site, including accessible trails, and a section of Sustrans Route 4. A poignant memorial to the 112 miners – men and boys – killed in a gas explosion in 1892 lies within the trees to the south of the site.

There is ample parking at a visitor centre (Access point), with good facilities. As the site is managed for wildlife, dogs should be kept on leads. Parc Slip is included in the Valleys Regional Park (see also Afan Forest Park [254], Cwmcarn Forest [320] and Dare Valley Park [268]).

COED Y FELIN

LLYS-FAEN/LISVANE, CARDIFF
MAP REF 278

Ownership: Public : Cardiff City Council
Designations:
Area: 7ha
Forest type: broadleaved
Forest location: ST181830
Explorer Map: 151
Ease of access: Easy
Access point: ST1830283098
///incomes.insect.branded
CF14 0TQ

Surrounded by housing, and lying close to the M4 motorway, this small woodland provides valuable habitat for wildlife and green lungs for local people. A circular nature trail can be followed, which includes a few steps, but keep an eye out for the dragon (sculpture!). About 20 bird boxes have been erected by local volunteers to support breeding birds. Not be confused with a wood of the same name [79] in the North-East region.

BRYNNA WOODS

BRYNNA, RHONDDA CYNON TAF
MAP REF 279

Ownership: Private: Wildlife Trust of South and West Wales
Designations:
Area: 26ha
Forest type: broadleaved
Forest location: SS990830
Explorer Map: 151
Ease of access: Easy
Access point: SS9853783089
///nesting.bloom.unique
CF72 9QJ

With housing on all four sides, with the main railway line forming its southern border, this site is an oasis for wildlife. The Afon Ewenni/River Ewenny and its tributaries dissect the woods, which, together with a sizeable pond, provide wonderful habitat for insects and their nocturnal predators, notably lesser horseshoe and barbastelle bats. Dormice live among stands of hazel, while an ancient oak tree thought to be 300 years old provides character and shelter for woodland birds. A small colliery was active on the site in the second half of the nineteenth century, with evidence of blocked off adits remaining under the trees. Near a large meadow to the east of the site is a bridge constructed to allow graziers to cross the now disused Ogmore–Cardiff railway line. There is no formal parking nearby and the Access point is a little hidden down a narrow road between a primary school and housing.

COED Y BEDW

PENTYRCH, CARDIFF
MAP REF 280

Ownership: Private: Wildlife Trust of South and West Wales
Designations: SSSI
Area: 17ha
Forest type: broadleaved
Forest location: ST110826
Explorer Map: 151
Ease of access: Moderate
Access point: ST1051582355
///issues.glare.pirate
CF15 9PQ

Along the valley bottom of the Nant Cwmllwydrew, acidic soils support alder carr and several scarce invertebrates, including giant lacewing, whose large (25mm) spotted wings make it quite distinctive. Its larvae live in the quiet waters, fringed with golden globes of marsh marigold. On the wooded slopes above, the lime-rich soils and springs mainly support ash trees accompanied by oak and birch, with hosts of wild garlic. The range of acid and alkaline soils across the site make it an unusual and interesting reserve.

Veteran broadcaster David Attenborough planted a beech tree here in 1985 when the reserve was opened, following its purchase from the Forestry Commission, although its precise whereabouts are already lost in trees and time. The reserve's development was supported by pioneering conservationist Mary Gillham (1921–2013).

The main access to the reserve lies to the east of the Access point along the busy minor road (ST1130882656 | ///beans.order.spend) but there is no parking nearby. Instead, park carefully near some field gates (Access point) and walk downhill following the public footpath across some fields to the valley bottom, before entering the wood. Paths are not well marked and often muddy.

COED LLWYN

LLANDDUNWYD/WELSH ST DONATS, VALE OF GLAMORGAN

MAP REF 281

Ownership: Private: Wildlife Trust of South and West Wales **Designations:** **Area:** 5ha	**Forest type:** mixed **Forest location:** ST040779 **Explorer Map:** 151 **Ease of access:** Easy/Moderate	**Access point:** ST0359178708 ///vine.elders.proceeds CF72 8JU

The Scots pine trees on this nature reserve support one of the largest heronries in Wales. Visitors may find the reserve closed due to dangerous ash trees, but the breeding colony, often more than 20 pairs, can be seen from a right of way (ST040781). The sight of a siege of herons (their collective noun) is spectacular. The adults arrive in late January and stay until their young fledge in July or August. It is also possible to reach the site from Hensol Forest [282], enjoying a pleasant walk along quiet country lanes.

HENSOL FOREST

LLANDDUNWYD/WELSH ST DONATS, VALE OF GLAMORGAN

MAP REF 282

Ownership: Public: Natural Resources Wales **Designations:** HPG, SSSI **Area:** 91ha	**Forest type:** mixed **Forest location:** ST034769 **Explorer Map:** 151 **Ease of access:** Easy/Moderate	**Access point:** ST0309576829 ///opened.pylon.bronzer CF71 7SW

Hensol Forest contains areas of conifers as well as pedunculate oak, ash, silver birch and rowan. The site has been used for TV productions, including the time-travelling BBC programme *Dr Who*, yet its own origins date back to 1600. Pysgodlyn Mawr, a lake surrounded by trees and designated an SSSI in the south-eastern corner of the forest (ST0426276088 | ///nametag.royal.bounded), harbours the rare medicinal leech. This is the largest species of leech (up to 20cm) found in Britain, and the only one that feasts on human blood. Prey includes other mammals, plus birds, fish and amphibians. With three jaws and 100 teeth, they are well-adapted to clinging onto their prey while they feed.

The heronry at Coed Llwyn [281] can be included as part of a circular route.

COED Y BWL

CASTLE-UPON-ALUN, VALE OF GLAMORGAN　　　**MAP REF 283**

Ownership: Private: Wildlife Trust of South and West Wales
Designations: SSSI
Area: 2ha
Forest type: broadleaved
Forest location: SS908749
Explorer Map: 151
Ease of access: Easy/Moderate
Access point: SS9096075085
///chat.edit.dwell
CF32 0TL

Once upon a time, the mighty English elm dominated this ancient woodland, growing alongside the Afon Alun/River Alyn, accompanied by ash, the queen of trees. After Dutch elm disease decimated the population of 30 million trees spread across the whole of Britain, it was ash that thrived in its place. Sadly, reflecting an increasing incidence of serious tree pests and diseases in Britain, ash is now declining due to a dieback disease caused by a non-native fungal pathogen. Some trees in this wood may yet recover, so allowing them time to do so is important, yet this presents a challenge for woodland owners due to risks to the public from falling branches, or even trees. Visitors may therefore find the nature reserve closed to the public. However, it is still possible to enjoy the ground flora and birdlife from the quiet lane along its lower edge, or to follow one of the public rights of way through the wood. The wood is famed for its wild daffodils, which almost outshine a stunning display of bluebells and wood anemones each spring, accompanied by the songs of blackcap, chiffchaff and willow warbler. Park in a small lay-by (Access point) near a pedestrian bridge over the river.

COED GARNLLWYD

LLANCARFAN, VALE OF GLAMORGAN　　　**MAP REF 284**

Ownership: Private: Wildlife Trust of South and West Wales
Designations: SSSI
Area: 14ha
Forest type: broadleaved
Forest location: ST058711
Explorer Map: 151
Ease of access: Moderate/Difficult
Access point: ST0517070467
///consults.triads.saved
CF62 3AH

Overlying a type of particularly hard limestone, known as lias, this small nature reserve supports a wonderful range of woodland plants that thrive on lime-rich soil. Below the ash and oak trees, many of which are old coppice, is a vibrant shrub layer comprising crab apple, field maple, hazel, hawthorn, holly, spindle and wayfaring tree. Ground flora confirms the ancient status of this wood, especially Goldilocks buttercup, herb-Paris and early purple orchid. Birdlife to spot includes lesser spotted woodpecker, nuthatch and treecreeper, while from November woodcock fly in from Scandinavia to overwinter. Look for the brown argus butterfly, and many interesting moths such as the scarce Blomer's rivulet.

There is no possibility of parking near the wood. Instead, park in Llancarfan Village, walk to the Access point and follow the public footpath up the valley to the north-east. Within the wood, the paths are steep and often muddy.

Crab apple.

Castle Ditches and wild clematis at Cwm Colhuw [286].

CWM COLHUW

LLANILLTUD FAWR/LLANTWIT MAJOR, VALE OF GLAMORGAN

MAP REF 286

Ownership: Private: Wildlife Trust of South and West Wales	**Forest type:** broadleaved	**Access point:** SS9569067484 ///recur.shadow.girder CF61 1RF
Designations: SM	**Forest location:** SS960675	
Area: 7ha	**Explorer Map:** 151	
	Ease of access: Moderate	

Behind the rocks and pebbles of Llantwit Beach – popular with families, fossil hunters and surfers – a narrow strip of ash woodland and scrub stretches up Cwm Colhuw, part of which is managed as a nature reserve. Look for bullfinch, linnet, whitethroat and willow warbler in the branches of the sea-stunted blackthorn trees that line the path. Three impressive earth banks run through the wood, known as 'Castle Ditches', once the defensive ditches for an Iron Age castle situated on the promontory overlooking the seas. Its steep slopes are covered by ramsons in spring. On its grassy plateau, look for green woodpecker feeding from the numerous ant hills, many butterflies (including ringlet and small blue), and keep an eye out for peregrine falcon hunting over the cliffs. The long-distance Valeways Millennium Heritage Trail (101km) passes through the scrubby woodland north of the Afon Colhuw. A stone stile at the north end of the wood (SS9674967836 | ///acids.shops.breezes) provides an alternative access point. The site is the most southerly featured in this guide.

South-Central 229

Trees grow on a regenerating old slag heap.

Green to black and back

The valleys of South Wales are synonymous with the coal mining industry, and also famed for its male voice choirs, strong trade unions and as a proud bastion of the Welsh language. Once within the historic county of Glamorgan, the valleys now fall within Rhondda Cynon Taf and Merthyr Tydfil. The main industrial valleys are the Rhondda (comprising two valleys: the larger Fawr valley and the smaller Rhondda Fach), Cwm Cynon/Cynon Valley and Ynysowen/Merthyr Vale.

Prior to the mid-nineteenth century, the valleys were remote and populated by farming communities. Soon after the first coal mines were sunk, the railway network began to expand, reaching the heads of the two Rhondda valleys at Treherbert (Fawr) and Maerdy (Fach) by 1856. Coal production peaked during both world wars, suffering a depression in between, with 15,000 miners employed and living in the Rhondda valley by 1947. The landscape of the valleys was utterly transformed from lush green to industrial black, littered with spoil heaps, chimneys, mine buildings and rows of terraced housing, and steeped with smoke pollution.

The valley bottoms were originally well-wooded, according to evidence from place names such as coed and gelli (copse), old maps, and the written accounts of Victorian travellers. As the mining industry developed, much of the original natural forest disappeared, whether simply cleared to make way for 'progress', or felled for its products including pit props (oak), charcoal production (alder) and firewood. Only the waning of the mining industry beckoned interest in forestry. The Forestry Commission started planting in the valleys in the 1960s, transforming the landscape again. The value of forestry and green space is well recognised today, especially its power to transform well-being and culture. Recently, investment in the valleys from social enterprises, energy firms and government is helping local people engage with their local environment.

A significant forested area in the valleys is managed as open access land by Natural Resources Wales and included as sites within the Valleys Regional Park. For the visitor there is much to explore, from fragments of native woodland and expansive conifer plantations, waterfalls and lakes, and of course remnants of industrial heritage. Former coal sites featured include Clydach Vale [273], Cwm Saerbren [270], Dare Valley Park [268], Gethin Woodland Park [267], Parc Slip [277], Parc Coedwigaeth Penpych [269] and St Gwynno Forest [272].

Trees are restoring beauty and nature to many sites once scarred by industrial mining.

PORTHKERRY WOODS

Y BARRI/BARRY, VALE OF GLAMORGAN

MAP REF 285

Ownership: Public: Vale of Glamorgan Council
Designations: LNR, SM, SSSI
Area: 64ha
Forest type: broadleaved
Forest location: ST093673
Explorer Map: 151
Ease of access: Easy/Moderate
Access point: ST0933167388
///certified.visitor.leaky
CF62 3BY

Three separate woods cloak the valleys west of Y Barri/Barry, managed by the local council as a country park and collectively known as Porthkerry Woods. To the south, and nearest the coast, lies the aptly named Cliff Wood. It runs alongside a housing estate before reaching steep limestone cliffs, and this is a good place to look for the endangered true service. This rare tree is distinguishable from the wild service tree by its compound pinnate leaflets growing on either side of the leaf's midrib (like a rowan or ash), as opposed to the single palmate (hand-like) leaf of its near relative. The wood's main canopy faces inland, comprising ash and pedunculate oak, with occasional field maple, holm oak and yew.

Mill Wood is nearer to the Access point and follows the steep-sided valley of Nant Talwg. A water-driven sawmill operated here from the mid-nineteenth century. Knockmandown Wood lies north of the main valley and is the least accessible of the three woods, but provides a wonderful silvan backdrop to the Porthkerry Viaduct (railway).

The whole site was used as a marshalling area for D-Day landing forces, with 4,000 troops and their equipment billeted here during 1944. A little more than a century earlier, Ann Jenkin, who lived in a cottage in Cliff Wood, was convicted of witchcraft (the remains of the cottage are still visible).

A choice of trails and paths, some of which are steep and include some uneven steps, can be followed around the site, including Lover's Lane, which passes through Cliff Wood. Parking is charged.

The arches of Porthkerry Viaduct at Porthkerry Woods [285]. Notice the steps (far right) climbing into Cliff Wood.

The South-East region incorporates the counties of Monmouthshire, Newport, Caerphilly, Blaenau Gwent and Torfaen, with the Bristol Channel at its southern border, and England to the east. It has below-average forest cover for Wales, with about 20,000ha (12.4 per cent forest cover), yet includes some of the most densely wooded areas in the country, along the spectacular Wye Valley. Sites include Bargain Wood [312], which inspired William Wordsworth, and Beacon Hill [310], Highmeadow Woods [298], Pentwyn Farm [300], Prisk Wood [301] and Wyndcliff Wood [315]. Like neighbouring South-Central, but at a lesser scale, this region also helped

Beautiful countryside surrounding the Punchbowl [299].

fuel the Industrial Revolution with its rich reserves of minerals and coal, as at Blaenavon Community Wood [304], Ebbw Vale Central Valley [303] and Llanfoist Wood [296]. Meanwhile, part of the Bannau Brycheiniog/Brecon Beacons National Park extends into the region's northern boundary. Sessile oak woods at Deri-Fach [292] grow high up on the slopes of Y Fâl/Sugar Loaf, meanwhile at nearby Bryn Arw [291] an inspiring organisation, Stump Up For Trees (see pp.240–41), is finding new ways to engage hill farmers with tree planting. The ancient hunting territory of Wentwood Forest [317] extends for more than 1,000ha, providing valuable space for nature and people.

SOUTH-EAST

MONMOUTHSHIRE, NEWPORT, CAERPHILLY, BLAENAU GWENT, TORFAEN

SITES 287–325

MYNYDD DU

PARTRISHOW, MONMOUTHSHIRE   MAP REF 287

Ownership: Public: Natural Resources Wales Designations: NP Area: 1,113ha	Forest type: conifer Forest location: SO251263 Explorer Map: OL13 Ease of access: Moderate/Difficult	Access point: SO2671525136 ///nozzles.rising.differ NP7 7NG

Mynydd Du is a large conifer plantation in the Y Mynyddoedd Duon/Black Mountains. It offers many kilometres of informal access, both for walkers and mountain bikers. From the Access point at Cadwgan car park, a lengthy circuit through the forest can be undertaken, exploring both sides of the pretty waters of the Grwyne Fawr. The high ridge to the west can be attained, leading up to the highest peak in the range, Waun Fach (811m). Another access option at the northern tip of the forest is Mynydd Du car park (SO2527028434 | ///choppers.flaking.impressed), offering better access for walkers wanting to explore the valley above the trees. Grwyne Fawr bothy (SO2340630670 | ///ownership.spine.overused) lies 2km upstream near the reservoir, surrounded by scattered rowan trees.

Offering a very different experience, lying between the two car parks is Nant-y-Bedd forest garden (SO2572726857 | ///fury.claps.inefficient). Open to the public under the National Garden Scheme, it includes an edible forest area, a two-storey treehouse with yoga platform, and a natural swimming pool, set among mixed trees including some tall Douglas fir. Entry is charged, pre-booking required.

STRAWBERRY COTTAGE WOOD

STANTON, MONMOUTHSHIRE   MAP REF 288

Ownership: Private: Gwent Wildlife Trust Designations: NP, SSSI Area: 6ha	Forest type: broadleaved Forest location: SO313215 Explorer Map: OL13 Ease of access: Moderate	Access point: SO3121121486 ///baguette.wardrobe.quilt NP7 7NB

Far below circling red kite and buzzard, the canopy of sessile oak trees in Strawberry Cottage Wood cloaks the steep hillside above the Afon Honddu. In spring, blackcap, pied flycatcher and redstart flit between their branches, looking down upon a woodland floor carpeted with bluebell, common dog violet, primrose and yellow archangel. The dormouse makes its home among the hazel understorey. From a small lay-by (Access point), walk across a pedestrian bridge over the river. If you are stealthy, you may be lucky enough to see an otter, while dipper, grey wagtail and kingfisher are common. Bryn Arw [291] can also be reached from here.

Yellow archangel grows among stitchwort and bluebell.

COED Y CERRIG

STANTON, MONMOUTHSHIRE

MAP REF 289

Ownership: Public: Natural Resources Wales
Designations: IPA, LNR, NNR, NP, SAC, SSSI

Area: 70ha
Forest type: broadleaved
Forest location: SO293211
Explorer Map: OL13

Ease of access: Easy/Moderate
Access point: SO2930521169
///protected.still.vegans
NP7 7NA

The value of this site for nature is evident from its multiple conservation designations (the NNR covers 10ha). Growing on both sides of the deep, steep-sided valley, once scooped out by glacial ice, the woodland is rich and varied. The valley bottom is mostly alder and willow carr (wet woodland), transforming to ash and hazel as the ground rises, with sessile oak and birch on higher and drier ground. Notable species among the ground flora include bird's-nest orchid, which being devoid of chlorophyll is therefore the colour of leaf litter and can be hard to spot! This rare orchid is saprophytic, meaning it relies on decaying plant materials for nutrients. Also present is the parasitic plant toothwort, which can be seen growing from hazel roots. Look also for nettle-leaved bellflower and herb-Paris under the trees, while marsh marigold, yellow pimpernel and the rare marsh fern grow in wet areas. The wood is alive with birdsong in spring and host to pied flycatcher, redstart and wood warbler. Silver-washed fritillary butterflies are among 13 species recorded on the site. The site supports a healthy population of dormice.

Toothwort.

From the small car park (Access point), a short easy circuit with sections of boardwalk passes through some of the carr woodland (south of the road). Another circular trail heads north into the wood, involving steps and some steep sections.

On the summit of the hill, with commanding views across the countryside, is the Iron Age hill fort Twyn-y-Gaer, but this is best approached via another route.

Marsh marigolds at Coed y Cerrig [289].

GRAIG SYFYRDDIN

CROSS ASH, MONMOUTHSHIRE   MAP REF 290

Ownership: Private **Designations:** **Area:** 88ha **Forest type:** mixed	**Forest location:** SO404216 **Explorer Map:** 189 **Ease of access:** Moderate	**Access point:** SO3957820924 ///brands.slick.lifeguard NP7 8LD

Many people consider Ysgyryd Fawr [293] an outlier of the Y Mynyddoedd Duon/Black Mountains, but further east still stands another prominent hill formed of the same old red sandstone. Graig Syfyrddin is a ridge with two summits, its southern peak is the highest, known colloquially as Edmund's Tump (423m). Although the top of the ridge is treeless, a conifer plantation wraps around half of it, with a broadleaved woodland below, mainly consisting of birch. A very satisfying short circular walk can be enjoyed, taking in the ridge, from where truly spectacular views can be enjoyed overlooking the Welsh Marches, the Black Mountains and even the distant Forest of Dean. A short segment of the challenging 32km-long Three Castles Walk passes through the trees, linking the Welsh castles of Grosmont, White and Skenfrith.

BUCKHOLT WOOD

TREFYNWY/MONMOUTH, MONMOUTHSHIRE   MAP REF 294

Ownership: Private **Designations:** SM **Area:** 68ha **Forest type:** mixed	**Forest location:** SO504162 **Explorer Map:** OL14 **Ease of access:** Moderate	**Access point:** SO5104015574 ///locating.cooks.earplugs NP25 5RE

A great number of public footpaths criss-cross this diverse wood, whose northern part lies within England. Many of these are steep and can be muddy. A prehistoric hill fort (SO5020815922 | ///engrossed.tripped.hotspot) is hidden among the trees, offering breathtaking views from its south-westerly escarpment looking across to the Afon Mynwy/River Monnow. Park in a small lay-by (Access point) or take advantage of a choice of public footpaths to enjoy a longer circular route starting from Trefynwy/Monmouth.

COED GWRAIG

TAL-Y-COED, MONMOUTHSHIRE   MAP REF 295

Ownership: Private: Woodland Trust **Designations:** **Area:** 4ha	**Forest type:** broadleaved **Forest location:** SO409155 **Explorer Map:** OL14 **Ease of access:** Moderate	**Access point:** SO4097215665 ///conducted.bland.stags NP7 8TL

Coed Gwraig is a small tranquil woodland, dominated by pedunculate oak but accompanied by a wide range of other tree species, including ash, birch, small-leaved lime, sycamore and wild service, while hazel, holly, spindle, dog rose and old coppiced hawthorn form a thick understorey. Another distinctive element of the site is the presence of many yew trees, which line its ancient sunken lanes. A good amount of deadwood provides wonderful habitat for invertebrates and fungi. From the gateway and small lay-by (Access point), several permitted paths can be followed through the wood.

Volunteers inspecting young trees at Bryn Arw [291].

BRYN ARW

BETTWS, MONMOUTHSHIRE   MAP REF 291

Ownership: Private: Bryn Arw Commoners Association	**Forest type:** broadleaved	**Access point:** SO2921920037
Designations: NP	**Forest location:** SO305198	///doll.blasted.cackling
Area: 73ha	**Explorer Map:** OL13	NP7 7LU
	Ease of access: Moderate/Difficult	

Bryn Arw (384m) is a long, narrow hill, isolated from others in the Y Mynyddoedd Duon/Black Mountains. Its steep (in places more than 30 degrees) east-facing slopes were used to great effect to launch Stump Up For Trees (see pp.240–41), when its name, cut out of the bracken in giant letters, was clearly visible from the A465. Since then, funding and permissions have been secured to fence the land to keep grazing livestock away, and 135,000 native broadleaved trees planted by local volunteers. Completed in March 2021, the site offers a wonderful opportunity for visitors to return over time to watch this young forest grow. Meanwhile, its summit offers spectacular views across the hedges, copses, dingles and woods of the Monmouthshire countryside, and the prominent summit of Y Fâl/Sugar Loaf to the west. In the future, giant letters will return to the slope once a year, when hundreds of hawthorns blossom each May to spell out 'Daw Eto Ddail Arfryn' (Leaves return to the hill).

Reaching the site from any of the possible entry points requires a good level of fitness and competency in map reading. The best access choices are either from the east (see Strawberry Cottage Wood [288]), or from the west at the National Trust car park at Fro (Access point).

South-East 239

Stump Up For Trees

As one of the most ambitious community-based organisations in Wales, Stump Up For Trees is aiming to plant at least one million trees across the hills and valleys of the Bannau Brycheiniog/Brecon Beacons. Among its many innovative approaches, the charity works closely with hill farmers using appropriate language and strong farming-sensitive propositions to help convince landowners of the benefits derived by planting more trees on their land. Its central proposition is that farmers' least-productive land could be allocated to trees, enhancing biodiversity, improving water flow and

sequestering carbon. Planting trees will not have a negative impact on their farming business, but will actually enhance it. While politicians and organisations may talk about 'public goods', and 'ecosystem services', Stump Up For Trees makes an accessible and compelling case that speaks with grounded authority to local landowners.

In 2020, the charity announced its intention in spectacular and unprecedented scale by partially mowing a bracken-clad hillside overlooking the A465, one of the main roads through the area, leaving some areas uncut to form giant letters. They spelt out 'Stump Up For Trees' on one line, and 'Daw Eto Ddail Arfryn' on the other, a Welsh phrase meaning 'Leaves return to the hill'. The hillside was one of the first sites acquired for tree planting in partnership with a local farmer. Known as Bryn Arw [291], more than 100,000 native trees have since been planted here. It was the first significant tree-planting project of this scale on common land anywhere in Wales.

Another innovative approach adopted by Stump Up For Trees has been its success in blending public and private finance to support tree planting. At Bryn Arw, the Welsh Government part-funded the scheme, while carbon credits were sold to a private company.

Much of the practical work, including tree planting, has been achieved by working with local people and other volunteers. The charity has also set up its own tree nursery to raise its own seedlings using seeds collected from local trees.

While fundraising to support more tree planting at scale, the charity now runs an education programme with local schools, helping shape future understanding and compassion for the environment.

Stump Up For Trees is investing a huge effort into recording wildlife on its sites by conducting surveys of birds, butterflies and vascular plants. Such long-term monitoring will help track the many positive impacts on the environment following its ambitious tree-planting activities across the beautiful landscape of the Bannau Brycheiniog.

Keith Powell (*left*) and Robert Penn (*right*) of Stump Up For Trees complete tree planting at Bryn Arw [291].

DERI-FACH

Y FENNI/ABERGAVENNY, MONMOUTHSHIRE   MAP REF 292

Ownership: Private: National Trust **Designations:** NP, SAC, SSSI **Area:** 69ha **Forest type:** broadleaved	**Forest location:** SO275170 **Explorer Map:** OL13 **Ease of access:** Easy/Moderate/Difficult	**Access point:** SO2684416725 ///hypocrite.sensitive.striving NP7 7EN

Three discrete blocks of trees can be found below the unmistakable summit of Y Fâl/Sugar Loaf (596m), together forming 120ha of internationally significant sessile oak woodland: The Park (east); Deri (south-east); and the largest of the three (south) in St Mary's Vale, known as Deri-Fach. These grow at the extreme south-eastern limit for sessile oak woods in Britain.

The sessile oak trees are accompanied by ash, alder, birch, hazel, holly and rowan. The north-east slopes in the vale are cool and shady, creating perfect conditions for bryophytes and ferns. One of Britain's largest leafy liverworts, the greater whipwort, thrives here. The south-western slopes are warmer, their sunny glades providing ideal conditions for ants, including red wood ant.

The Access point is a car park popular with walkers, offering spectacular views across the Usk Valley and far beyond. The wood can be reached by following any of the many paths that criss-cross the moorland.

A visitor aiming to explore all three sessile oak woods in a day would need to be a competent navigator and possess a good level of fitness, starting at an alternative car park nearer the centre of the group (SO2888016636 | ///dress.waggled.apparatus).

The trees of Deri-Fach [292].

LLANFOIST WOOD

LLAN-FFWYST/LLANFOIST, MONMOUTHSHIRE   MAP REF 296

Ownership: Private **Designations:** NP, SM, World Heritage Site **Area:** 12ha	**Forest type:** broadleaved **Forest location:** SO287126 **Explorer Map:** OL13 **Ease of access:** Easy/Moderate	**Access point:** SO2861113302 ///doormat.listening.foggy NP7 9YA

Running alongside the Monmouthshire & Brecon Canal, and on the lower slopes of Blorens Hill, this attractive woodland provides a perfect accessible green space near to the town of Llan-ffwyst. It offers opportunities to observe common woodland birds, spot fungi in the autumn and to watch life passing by on the canal.

From the public car park (Access point), walk south, following signs pointing towards Sustrans Route 49, which is reached up the short hill just before the canal. To access the main body of the wood, carry straight on (i.e. not along Route 49), passing through a small tunnel under the canal. An easy circuit can be followed by walking through the wood, and back along the towpath on the other side of the canal (also Route 49). A longer circular route using public footpaths returns via the top of the wood, finally descending along the old tramroad and incline built in 1818 to carry coal, iron and limestone to a wharf on the canal. Various archaeological remains can be discovered, including the stone sleeper blocks that once supported the tram rails. The site falls within the Blaenavon Industrial Landscape World Heritage Site (see also Blaenavon Community Wood [304]). A keen walker could also extend the longer route to visit nearby Punchbowl [299].

The Monmouthshire & Brecon Canal.

YSGYRYD FAWR

Y FENNI/ABERGAVENNY, MONMOUTHSHIRE   MAP REF 293

Ownership: Private: National Trust Designations: NP Area: 45ha Forest type: mixed	Forest location: SO326178 Explorer Map: OL13 Ease of access: Moderate	Access point: SO3290816425 ///tablet.begun.rotations NP7 8NL

Ysgyryd Fawr (486m), also known as The Skirrid, is a prominent outlying hill of the Y Mynyddoedd Duon/Black Mountains. It is a popular destination for walkers, most of whom head for the summit, climbing up through Caer Wood at the start of the ascent. At the point where most walkers continue upwards for open ground, an enjoyable alternative can be followed west and then north through the ash, sycamore and hazel trees that cloak the western slopes of the hill. The path contours around the hill, and though it can be muddy in places, boardwalks cover the wettest sections. At the north end of the hill, it is possible to continue with a circumnavigation of the hill, though the eastern flanks are almost treeless. Alternatively, a steep ascent can be made to the summit and the remains of St Michael's Chapel. The views in every direction are wondrous: north to the Malvern Hills in Herefordshire; east to the Forest of Dean and another outlying hill, Graig Syfyrddin [290]; south to the Usk Valley and Somerset; and west to the Y Mynyddoedd Duon/Black Mountains and the Bannau Brycheiniog/Brecon Beacons. The Access point is a large car park (charge).

Looking towards Y Fâl/Sugar Loaf from Ysgyryd Fawr/The Skirrid [293].

The distinctive silhouette of Ysgyryd Fawr/The Skirrid [293].

244   The Forest Guide: Wales

BEAULIEU WOOD

TREFYNWY/MONMOUTH, MONMOUTHSHIRE — MAP REF 297

Ownership: Private: Woodland Trust **Designations:** NL **Area:** 17ha	**Forest type:** broadleaved **Forest location:** SO526128 **Explorer Map:** OL14 **Ease of access:** Moderate	**Access point:** SO5230512926 ///tolerates.rebel.putter NP25 3SD

On the slopes below the eighteenth-century buildings, the Round House and Naval Temple, collectively known as The Kymin (owned by the National Trust), the ancient woodland site being restored at Beaulieu is well worth a visit. Characterful beech, birch and oak trees, released from conifers planted in the 1950s, grow between giant moss-clad boulders. Climb up from Monmouth town along the Offa's Dyke Path to reach the Access point (no parking), or walk down from The Kymin.

HIGHMEADOW WOODS

TREFYNWY/MONMOUTH, MONMOUTHSHIRE   MAP REF 298

Ownership: Public: Natural Resources Wales
Designations: NL, NNR, SAC, SSSI
Area: 810ha
Forest type: mixed
Forest location: SO542139
Explorer Map: OL14
Ease of access: Moderate
Access point: SO5411012512
///radiates.turkey.gourmet
NP25 3SB

Highmeadow Woods is a large area of mixed woodland to the west of the Forest of Dean, forming part of Redding's Inclosure. At its north-western fringes is Lady Park Wood, an NNR managed with minimal intervention and valued especially for its deadwood habitat, rare plants, breeding birds and bats. Ash, beech, lime, oak and yew thrive on the limestone bedrock, supporting a rich understorey of hazel, holly and honeysuckle. The reserve is a fragile and important wildlife site protected by a high deer fence designed to exclude both deer and people, and should not be accessed by the public.

The Wysis Way threads its way through Highmeadow Woods, offering spectacular scenery including Britain's largest boulder, the Suck Stone. Close by are the crags of Near Hearkening Rock (SO5426613997 | ///prompts.asked.obstruct) from where dramatic views can be enjoyed across Wales, taking in the Wye Valley and the distant mountains of the Bannau Brycheiniog/Brecon Beacons and Y Mynyddoedd Duon/Black Mountains.

The Access point lies within England, as does a substantial part of Highmeadow Woods, but most of the more interesting trails and features are found within Wales. An alternative route for those who enjoy a challenge is to start in Trefynwy/Monmouth town and follow the Wye Valley Walk north as far as the suspension footbridge at Biblins. There the river can be crossed before climbing the steep path directly up the hill, which follows the fenced boundary of Lady Park Wood. Beaulieu Wood [297] and The Kymin provide a very satisfying conclusion towards the end of a quite demanding yet very scenic 13km circular walk.

Views across the canopy of Highmeadow Woods [298] from the crags of Near Hearkening Rock.

Pentwyn Farm [300].

PENTWYN FARM

PEN-TWYN, MONMOUTHSHIRE — MAP REF 300

Ownership: Private: Gwent Wildlife Trust Designations: NL Area: 1ha	Forest type: broadleaved Forest location: SO523093 Explorer Map: OL14 Ease of access: Easy	Access point: SO5236309385 ///caramel.finer.thudding NP25 4SA

The grassland areas of this nature reserve are designated an SSSI due to their rich flora, especially vibrant populations of orchids, including greater butterfly, green-winged and twayblade. The few existing copses of ash, oak and ancient small-leaved lime pollards have been linked together with new hedgerows, with added clumps of broadleaved trees across the open landscape, improving habitat for dormouse. Enjoy wonderful views across the Wye Valley towards The Kymin (see Beaulieu Wood [297]). The Access point is found at the end of a long gravel track, passable by car.

THE WERN

MITCHEL TROY, MONMOUTHSHIRE — MAP REF 302

Ownership: Private: Gwent Wildlife Trust Designations: NL Area: 3ha	Forest type: mixed Forest location: SO484087 Explorer Map: OL14 Ease of access: Moderate	Access point: SO4858408786 ///refutes.glove.prevented NP25 4JT

High above the Afon Troddi/River Trothy, which winds its way to the town of Trefynwy/Monmouth, this small nature reserve features scattered trees of ash, silver birch, holly and yew, interspersed with rough grassland and open heath. Adder and common lizard might be seen basking in the sun, especially near the shelter of the many old stone walls. Every surface is clad in mosses, while liverworts and ferns thrive under the trees and among the walls and millstone boulders.

PRISK WOOD

PEN-TWYN, MONMOUTHSHIRE

MAP REF 301

Ownership: Private: Gwent Wildlife Trust Designations: NL, SAC, SSSI Area: 6ha	Forest type: broadleaved Forest location: SO532089 Explorer Map: OL14 Ease of access: Moderate/Difficult	Access point: SO5318309142 ///waxes.jogged.happening NP25 4AL

Prisk Wood is not an easy site to reach or explore, which is partly why it is unspoilt and rich with wildlife. For several centuries from the medieval period onwards, these steep east-facing slopes rising high above the Afon Wye were busy, echoing with quarrying activities. The bedrock here is a quartz conglomerate, known as 'puddingstone', which was highly prized for millstones, especially for cider and perry mills. Formed from old riverbeds, its distinctive mix of rounded gravel was compressed together with fine sands and silts to form a natural concrete, its mottled appearance bearing a strong resemblance to Christmas pudding.

The remains of 20 quarries exist under the canopy of ash and small-leaved lime trees, and among the deep leaf litter is the occasional abandoned millstone. The exposed rock and mossy tree trunks harbour liverworts and ferns. Pied and spotted flycatchers visit in spring, while all three species of woodpecker can be seen. Bluebell and ramsons light up the wood in spring.

Apple and pear orchards once lined Lone Lane, from which the nature reserve can be accessed. Look for a yellow grit bin next to the entrance (Access point), although there is no room to park here. Instead, park at nearby Pentwyn Farm [300] and walk down the lane. If enjoying a longer ramble following the Wye Valley Walk, the wood can be reached by climbing up the hill from the valley bottom.

EBBW VALE CENTRAL VALLEY

GLYNEBWY/EBBW VALE, BLAENAU GWENT

MAP REF 303

Ownership: Public: Blaenau Gwent County Borough Council Designations: Area: 6ha	Forest type: broadleaved Forest location: SO172086 Explorer Map: 166 Ease of access: Easy	Access point: SO1736908522 ///zips.just.coffee NP23 6AN

The ex-industrial landscape of Glynebwy/Ebbw Vale has been transformed since the largest steelworks in Europe sited here closed in 2002. Today, a green corridor runs through the town, managed by the Gwent Wildlife Trust, featuring planted woodland, together with areas landscaped with wildflower-rich grassland, ponds and reedbeds. The valley provides a green oasis for local people and wildlife. Look for barn owl and kestrel hunting for voles in the long grass. In summer, grayling and dark green fritillary are among 10 or more species of butterfly to be found. Park next to the Environmental Resource Centre (Access point) and head to the site using the adjacent raised pedestrian walkway over the A4281 and railway line. Sustrans Route 466 passes through the valley.

PUNCHBOWL

LLAN-FFWYST/LLANFOIST, MONMOUTHSHIRE  MAP REF 299

Ownership: Private: Woodland Trust	Forest type: broadleaved	Access point: SO 2784911228
Designations: NP	Forest location: grid	///seagulls.lamp.seats
Area: 38ha	Explorer Map: OL13	NP7 9LD
	Ease of access: Moderate	

Long ago, on the shadowed east side of Blorens (555m), ice scooped out a large cwm from the hillside. Its crescent-shaped bowl is now lined with ancient woodland, with a small humanmade lake at its centre. Old beech pollards attest to former management as wood pasture, while the mature woodland that eventually took hold is now composed of sessile oak, ash, downy birch and rowan, with occasional holly and field maple. Look for raven riding the turbulent currents rising above the steep slopes, and chaffinch, robin and wren under the tree canopy. The tree-lined cwm is visible itself from far and wide, so it is no surprise that from the site, views across the Usk valley are spectacular. There is room for a few cars to park near a field gate leading into the site (Access point), where a stone wall encircles the remains of a lodgepole pine plantation.

The cwm of the Punchbowl [299] and the Usk Valley. The distinctive silhouette of Ysgyryd Fawr/The Skirrid [293] dominates the skyline.

BLAENAVON COMMUNITY WOOD

BLAENAVON, TORFAEN MAP REF 304

Ownership: Public: Natural Resources Wales **Designations:** SM, World Heritage Site	**Area:** 83ha **Forest type:** mixed **Forest location:** SO270074 **Explorer Map:** OL13	**Ease of access:** Moderate **Access point:** SO2648408269 ///blush.flirts.zips NP4 9HH

This woodland was once dominated by productive larch, but the trees were felled after the outbreak of *Phytophthora ramorum*. It has now been adopted by local people as a community wood. Areas of mixed conifers and broadleaves are successfully regenerating alongside spruce and some old, gnarly oak trees. An area of old coppiced beech – in its own right unusual because beech is more usually pollarded (cut at head height) – has earned the moniker of 'enchanted forest' as a result of the strange wavy multi-stemmed trees.

The remains of Capel Newydd, a late medieval chapel, can be discovered near to an alternative entrance further east (SO2705607708 | ///deflated.genius.evaporate). Old quarry workings also lie hidden under the trees.

The wood is within the Blaenavon Industrial Landscape World Heritage Site (see also Llanfoist Wood [296]). The region was of critical importance for coal mining and iron making during the late eighteenth and the early nineteenth centuries, fuelling the Industrial Revolution. Nearby Big Pit National Coal Museum is well worth a visit.

MARGARET'S WOOD

WHITEBROOK, MONMOUTHSHIRE MAP REF 305

Ownership: Private: Gwent Wildlife Trust **Designations:** NL **Area:** 2ha	**Forest type:** broadleaved **Forest location:** SO525069 **Explorer Map:** OL14 **Ease of access:** Moderate	**Access point:** SO5255107008 ///chopper.dining.outgrown NP25 4TX

Once a small farmstead dating from the eighteenth century, the former fields and orchard have since been swallowed by woodland. There is just room for one car in a lay-by at the reserve entrance (Access point), from where a path climbs up the steep hillside, passing close by the ruins of the farm cottage. The woodland of ash and hazel, with occasional holly, is unspoilt and a haven for wildlife. Visit in spring to see the nodding heads of wild daffodil, which grows extensively across the site. Manor Wood [306] lies nearby to the south.

Holly flowers.

MANOR WOOD

THE NARTH, MONMOUTHSHIRE   **MAP REF 306**

Ownership: Public: Natural Resources Wales Designations: NL, SM Area: 57ha	Forest type: mixed Forest location: SO528061 Explorer Map: OL14 Ease of access: Moderate	Access point: SO5234106289 ///universal.watches.gums NP25 4QW

This tranquil wood was once a place of industry. Old saw pits and charcoal manufacturing remains are scattered under the trees, while to the north-east near Whitebrook there was a paperworks, a grist mill (for grinding corn) and a cider mill. A seventeenth-century wireworks used the power of the Whitebrook to power its forge. Nearby today is a viewpoint offering an attractive view across the Wye Valley.

Park in a lay-by opposite the village hall at The Narth. Alternatively, there is another car park to the south (SO5290705884 | ///cooked.encourage.distilled). Exploring the wood from here requires more climbing, but it provides another access point to nearby Beacon Hill [310]. Whichever route is followed through the wood, the Manor Brook requires fording or leaping several times, and there are some steep sections with steps.

CROES ROBERT WOOD

TRELLECH, MONMOUTHSHIRE   **MAP REF 307**

Ownership: Private: Gwent Wildlife Trust Designations: NL, SSSI Area: 15ha	Forest type: broadleaved Forest location: SO478059 Explorer Map: OL14 Ease of access: Moderate	Access point: SO4757005982 ///pirates.hoops.donates NP25 4PL

An ancient woodland site without any ancient trees (they were felled in 1982), this small woodland is nonetheless a wonderful haven for wildlife. Dormouse thrives under the hazel coppice, which is managed on a traditional rotation of 5–8 years, and used to manufacture charcoal. The coppice management allows good amounts of sunlight to reach the ground where bluebell, lesser celandine and wood anemone burst spectacularly into flower each spring, followed by common spotted orchid. Several streams flow north through the site, lined by moisture-loving plants, including golden saxifrage. Listen for blackcap, willow warbler and great spotted woodpecker. Sparrowhawk and its larger relative the goshawk hunt through the ash canopy. Deadwood is being allowed to accumulate, sustaining a rich diversity of fungi and invertebrates. A network of paths criss-cross the site, and a circular trail can be followed. Paths are often muddy and steep in places.

Wood anemone.

PRIORY WOOD

## BETTWS NEWYDD, MONMOUTHSHIRE	MAP REF 308

Ownership: Private: Gwent Wildlife Trust **Designations:** SSSI **Area:** 5ha	**Forest type:** broadleaved **Forest location:** SO353058 **Explorer Map:** 152 or OL13 **Ease of access:** Moderate	**Access point:** SO3516605835 ///tour.used.shunning NP15 1JZ

Noctule bats thrive on a rich diet of invertebrates (more than 283 recorded) in this unspoilt ancient woodland. Its canopy is mostly oak, accompanied by ash, beech, silver birch and a stand of wild cherry towards the centre of the wood. Hawfinch might be seen feeding on cherry fruit in July, and the site is rich with other woodland birds all year round. Among the many wildflowers growing under the trees in springtime, look for bluebell, primrose, ramsons and wood sorrel. If approaching the site from the west, the road crosses the Afon Wysg/River Usk via Chain Bridge. The reserve entrance (Access point) is opposite an old brick pill box where there is room for a couple of cars to park. Paths through the wood can be steep and include some sections with steps.

Beech and holly at Priory Wood [308].

SILENT VALLEY

CWM, BLAENAU GWENT　　　　　　　　　　MAP REF 309

Ownership: Private: Gwent Wildlife Trust Designations: LNR, SSSI Area: 40ha	Forest type: broadleaved Forest location: SO187063 Explorer Map: 166 Ease of access: Moderate	Access point: SO1868705971 ///mining.steep.grafted NP23 7RX

Local lore suggests this nature reserve gained its name from steel workers looking for tranquillity and solace from their cacophonous industry at the nearby steelworks of Glynebwy/Ebbw Vale (see Ebbw Vale Central Valley [303]). Most trees lining the valley of Nant Merddog are beech, which, growing between 250m and 300m above sea level, make this one of the highest-elevation naturalised beech woodlands in Britain. Fungi abound in autumn, while great spotted woodpecker, nuthatch and treecreeper are among many woodland birds to be regularly seen. Some paths are muddy and wet underfoot throughout the year, although there are some short sections of boardwalk. The damp conditions favour snails and slugs, including Britain's largest, the ash-black slug, which can grow to 15cm long by feeding on fungi. An old tramway built to transport iron ore runs along the wood's upper boundary, which can be followed to make a circular route around the wood, offering good views across the valley.

BEACON HILL

TRELLECH, MONMOUTHSHIRE　　　　　　　MAP REF 310

Ownership: Public: Natural Resources Wales Designations: NL Area: 324ha	Forest type: mixed Forest location: SO512053 Explorer Map: OL14 Ease of access: Easy/Moderate	Access point: SO5103105230 ///dive.tailwind.mulls NP25 4QB

Lying west of Bigs Weir, the tidal limit of the River Wye, Beacon Hill (306m) offers commanding views over the Wye Valley. On its lower slopes are the aptly named Cuckoo Wood and the dramatic Duchess Ride viewpoint. To the south is Bargain Wood [312], which can be included as part of a longer ramble, following the Wye Valley Walk. There are also wonderful views westward to the Bannau Brycheiniog/Brecon Beacons and the distinctive outlines of Blorens, Ysgyryd Fawr/The Skirrid and Y Fâl/Sugar Loaf. The top of the hill and viewpoint are only a short walk from the car park (Access point), from where two waymarked trails can be followed. There are some magnificent mature Scots pine trees, while the heathland being restored across the site is rich with wildlife. After watching the sun set behind the mountains, linger a while to see nightjar emerge at dusk to feed on moths.

Scots pine.

LASGARN WOOD

ABERSYCHAN, TORFAEN

MAP REF 311

| Ownership: Public: Torfaen County Borough Council
Designations: SM
Area: 120ha | Forest type: broadleaved
Forest location: SO274040
Explorer Map: 152
Ease of access: Moderate | Access point: SO2709404207
///ocean.care.joyously
NP4 8XG |

Trees have reclaimed the hillside, which rises steeply on the east side of Abersychan. Once industry reigned here, including numerous small limestone quarries, supported by the Abersychan Limestone Railway constructed in 1826. While most quarrying and mining in the region at that time was connected with tramroads, this railway was pioneering new technology, adopting cast-iron 'fish-belly' rails (named after their distinctive cross-section). At the north end of the wood, a disused reservoir that once provided drinking water to the town is slowly being reclaimed by nature. A small car park (Access point) can be found opposite Victoria Village, just across the bridge over the Afon Lwyd. The path climbs steeply into the wood, which lights up in glorious autumn gold thanks to its beech trees.

BARGAIN WOOD

LLANDOGO, MONMOUTHSHIRE

MAP REF 312

| Ownership: Public: Natural Resources Wales
Designations: NL
Area: 53ha | Forest type: mixed
Forest location: SO524031
Explorer Map: OL14
Ease of access: Easy | Access point: SO5229402943
///active.thud.variances
P16 6NQ |

Bargain Wood, named Whitestone on visitor signage, provides one of the best views across the Wye Valley. The wood is mostly beech and oak with holly understorey, plus areas of conifers including Douglas fir, Norway spruce and western red cedar. A short circular route partly follows the brow of the hill, including a section of the Wye Valley Walk, passing a number of spectacular viewpoints. In 1798, William Wordsworth wrote his poem 'Lines Composed a Few Miles above Tintern Abbey' from the highest viewpoint, including the verse:

> O sylvan Wye! thou wanderer thro' the woods,
> How often has my spirit turned to thee!

The Hudnalls (a site in England) is visible on the opposite bank of the Wye, and in the middle distance, Bigsweir Wood (England) below Wyegate Hill.

Follow the Wye Valley Walk northwards to reach Cleddon Falls, which are especially dramatic following rainfall, while beyond lies Cuckoo Wood and Beacon Hill [310].

There is room for a few cars near the entrance to the wood (Access point), plus a larger car park further up the track next to a children's playground and picnic area.

Opposite: Beech and oak with holly understorey at Bargain Wood [312].

WYNDCLIFF WOOD

CAS-GWENT/CHEPSTOW, MONMOUTHSHIRE — MAP REF 315

Ownership: Public: Natural Resources Wales **Designations:** HPG, NL, SAC, SM, SSSI **Area:** 106ha	**Forest type:** broadleaved **Forest location:** ST526971 **Explorer Map:** OL14 **Ease of access:** Moderate/Difficult	**Access point:** ST5267697174 ///lighters.planet.pirates NP16 6HD

The woods at Wyndcliff fall under the Wye Valley Area of Outstanding Natural Beauty, and a Special Area of Conservation, which includes 16 separate SSSIs. These beautiful broadleaved woods, growing on steep slopes above and below the A466 (where it is known as 'Lower Wyndcliff Wood'), are mostly beech with occasional lime, ash and yew. The hazel understorey provides precious habitat for the dormouse, which thrives here. The woods are also known for the many rare species of *Sorbus*, including round-leaved whitebeam. Some 200m east of the car park (Access point) is a large limestone quarry, popular with climbers, harbouring two very rare tree species: English whitebeam and grey-leaved whitebeam.

The main attraction for most visitors, who may not notice the rare trees or

mammals, is a spectacular viewpoint constructed in 1828 for the Duke of Beaufort. The Eagle's Nest overlooks the sweeping meanders of the River Wye, the Piercefield Cliffs, and beyond to the Severn estuary and its bridges. Keep an eye out for peregrine falcon soaring above the cliffs. From the Access point a circular route can be followed to the Eagle's Nest by climbing 365 steps, many of which are uneven, and with occasional handrails where drops to the side are especially precipitous. Choose an anticlockwise direction to ascend the steps first and return via a steep path. There is an alternative car park nearer the top providing easier access, but the one detailed provides more options to explore the wood fully.

In Lower Wyndcliff Wood, the remains of an eighteenth-century plunge pool are waiting to be discovered (ST5248596819 | ///graceful.mealtime.dividers), another viewpoint at Lover's Leap, and beyond Giant's Cave (ST5246496281 | ///fairy.producers.downward) and Otter's Cave (ST5248696153 | ///redefined.usages.dripping). To explore Lower Wyndcliff Wood and these features, follow the waymarkers of the Wye Valley Walk south from the Access point along a relatively easy path, or approach from the south, starting at Cas-Gwent/Chepstow, passing through Piercefield Woods [318].

Looking out across the meanders of the River Wye from the Eagle's Nest in Wyndcliff Wood [315]. Both major road bridges across the River Severn are visible in the far distance.

TRANCH WOOD

PONTNEWYNYDD, TORFAEN   MAP REF 313

Ownership: Private: Gwent Wildlife Trust **Designations:** **Area:** 13ha	**Forest type:** mixed **Forest location:** SO269013 **Explorer Map:** 152 **Ease of access:** Moderate	**Access point:** SO2754201633 ///displays. mandolin.strategy NP4 6NE

Overlooking the town of Pontnewynydd, Tranch Wood incorporates part of the Branches Fork Meadows nature reserve. A good range of birds can be seen in this surprisingly tranquil peri-urban site. From the car park on Marchant's Hill (Access point), walk or cycle west along Sustrans Route 492 to reach the entrance to the nature reserve (SO2688701572 | ///saints.retail.fortunate). A short unsurfaced circular trail can be followed through the wildflower meadows and trees of the reserve, or extended to explore more of the woodland covering the slopes above.

COED-CWNWR

BRYNBUGA/USK, MONMOUTHSHIRE   MAP REF 314

Ownership: Private: Gwent Wildlife Trust **Designations:** SSSI **Area:** 17ha	**Forest type:** broadleaved **Forest location:** ST406995 **Explorer Map:** 152 **Ease of access:** Moderate	**Access point:** ST4106799157 ///grape.couple.poets NP15 1LL

Providing one of many valuable habitats at Springfield Farm nature reserve, the ancient woodland of Coed-Cwnwr is rich with wildlife. The working farm also includes flower-rich meadows, managed with livestock grazing and by haycutting, connected by hedgerows that provide corridors for birds and mammals to move across the site. Look for blackcap, chiffchaff and willow warbler along the hedgerows and woodland edges. Bluebells, sweet woodruff and ramsons carpet the woodland floor in spring, while the scarce adder's-tongue fern and broad-leaved helleborine can be found in shady areas alongside woodland rides. Several damp and shady valleys with trickling streams run through the wooded areas, including Llewelyn's Dingle at the southern boundary of the reserve. Paths criss-cross the reserve, allowing ample opportunity to enjoy spectacular views of the Usk Valley and the distant Y Mynyddoedd Duon/Black Mountains.

Sheep shelter under an old hedge.

The limestone Piercefield Cliffs.

PIERCEFIELD WOODS

CAS-GWENT/CHEPSTOW, MONMOUTHSHIRE — MAP REF 318

Ownership: Private: Gwent Wildlife Trust	**Forest type:** broadleaved	**Access point:** ST5288394334
Designations: NL, SAC, SM, SSSI	**Forest location:** ST533955	///pixies.went.mailboxes
Area: 82ha	**Explorer Map:** 167	NP16 5NZ
	Ease of access: Moderate	

More than 3km of beautiful woods stretch northwards from Cas-Gwent/Chepstow, lining the Wye Valley. They grow above the deeply incised meander of the river, marked by the dramatic limestone Piercefield Cliffs. The remains of two Iron Age hill forts are sited above, once commanding extensive defensive views across the valley but now clad in trees.

The woods comprise ash, field maple and wild service where soils are alkaline (above limestone rock), plus specimens of very rare *Sorbus* species, including English whitebeam. On acid soils, oak is more dominant with silver birch, rowan and beech. Small-leaved and large-leaved lime trees are also present, and home to the endangered bast bark beetle, a small and inconspicuous brown invertebrate found at only five sites in Britain. Scattered ancient yew trees are waiting to be discovered, some thought to be at least 750 years old.

From Cas-Gwent/Chepstow, follow the Wye Valley Walk northwards. Parking is available at Chepstow Leisure Centre (Access point). The route passes by The Alcove, an eighteenth-century folly, and further upstream, Otter's Cave, Giant's Cave and Lover's Leap, before reaching Wyndcliff Wood [315] and the Eagle's Nest. These features once formed part of the Piercefield Estate and its 'picturesque' landscape. A two-day tour of the Wye Valley and the 'Piercefield Walk' became highly fashionable after William Gilpin's guidebook *Observations on the River Wye* was published in 1782. Georgian tourists would halt at predestined locations to admire the scenery, commonly by turning their back on the landscape and viewing it through a small, convex and black-tinted mirror known as a Claude Glass. This toned down the contrast of a scene and compressed the image, making it appear more painterly or 'picturesque'.

WENTWOOD FOREST

LLANVAIR DISCOED, NEWPORT MAP REF 317

Ownership: Private: Woodland Trust, and Public: Natural Resources Wales Designations:	Area: 1,052ha Forest type: mixed Forest location: ST422948 Explorer Map: 152 and OL14	Ease of access: Easy/Moderate Access point: ST4221394877 ///roadways.chitchat.turkeys NP15 1NA

Wentwood Forest was a large ancient hunting forest, once the domain of the lords and earls of Chepstow Castle. This would have made it among the largest areas of ancient woodland in Wales, yet it has few old trees remaining after most of the native trees were felled during the Second World War and replaced with conifers. So it is an ancient woodland site, but no longer a forest of ancient trees. It is actively managed for timber production.

Today, its area is split between the modern counties of Newport and Monmouthshire, while ownership is split 50:50 between public and private landowners. The conifers are being progressively felled and either replaced by planting native trees or allowing natural regeneration with a range of species, both native and exotic. Two notable trees can be visited. Lying next to one of the popular paths is the Curley Oak (ST4043093680 | ///cement.meals.cobbles), an ancient tree with a girth exceeding 6m. It is crowded in by the straight stems of conifer trees, only accentuating its gnarly trunk, which is split completely through. Nearby stands an ancient pollard beech tree (ST4067093750 | ///dreading.stalemate.conductor).

There is a great deal of wildlife to see and hear, while more elusive species like dormouse may leave behind tell-tale signs (see p.179). Visit early in the morning to see fallow deer, and at dusk for a chance to see badger and bats (brown long-eared and pipistrelle). The large, mounded nests of southern wood ants are unmistakable, and attractive to green woodpeckers. The forest is a good place to spot fungi in autumn.

Left: Buzzard.

Below: Wentwood Forest and reservoir.

262 The Forest Guide: Wales

There are two main car parks; Cadeira Beeches (Access point) at the centre of the forest next to an avenue of beech trees, and Foresters' Oaks (ST4286193901 | ///provoking.upsetting.crowns) on the southern periphery near Wentwood Reservoir. A range of waymarked trails can be followed, some of which offer elevated views across to the Severn estuary.

COED PEN-Y-LAN

COED-Y-PAEN, MONMOUTHSHIRE

MAP REF 316

Ownership: Private **Designations:** **Area:** 4ha **Forest type:** mixed	**Forest location:** ST341975 **Explorer Map:** 152 **Ease of access:** Easy/Moderate	**Access point:** ST3414097529 ///televise.brands.prompting NP4 0TQ

Part of a larger (35ha) wood called The Forest, which belongs to multiple private owners, Pen-y-Lan is the area growing on the north-west escarpment bisected by a public footpath. The mostly broadleaved trees of ash, sweet chestnut and wild cherry are accompanied by vigorously regenerating birch and hazel, with a parcel of mature Norway spruce. The southern edge is lined with veteran beech trees and the rich ground flora hints at its ancient woodland origins. The Tithe Map shows that 'Forest Beeches' was an intensively worked woodland in 1838. Despite its proximity to Cardiff, Cwmbrân/Cwmbran and Newport, this is a quiet and secluded woodland, reached via an unsurfaced road.

COED MEYRIC MOEL

CWMBRÂN/CWMBRAN, TORFAEN   MAP REF 319

Ownership: Private: Gwent Wildlife Trust, and Public: Torfaen County Borough Council **Designations:**	**Area:** 8ha **Forest type:** broadleaved **Forest location:** ST272939 **Explorer Map:** 152	**Ease of access:** Easy **Access point:** ST2709794151 ///edits.sobs.swung NP44 6UT

Surrounded on all sides by housing, Coed Meyric Moel is a fragment of ancient woodland growing alongside a traditionally managed hay meadow. Running to the north of the wood is an old tramway and incline, linking Henllys Colliery with the Monmouthshire & Brecon Canal. Many woodland birds live and breed among the ash, beech and pedunculate oak trees, including blackcap, coal tit, great spotted woodpecker, nuthatch and treecreeper. Look for the flowers of yellow pimpernel among the dappled shade. The nature reserve covers a small proportion (1ha) of this now-urban wood. Park along the roadside of Tramway Close and enjoy surfaced routes through the trees.

CWMCARN FOREST

CROSSKEYS, CAERPHILLY   MAP REF 320

Ownership: Public: Natural Resources Wales **Designations:** SM **Area:** 350ha	**Forest type:** conifer **Forest location:** ST234931 **Explorer Map:** 152 **Ease of access:** Easy/Moderate	**Access point:** ST2284093618 ///spilling.preparing.ideals NP11 7FA

Cwmcarn Forest is popular with walkers and cyclists, offering an extensive network of waymarked trails and informal paths. The site is included within the Valleys Regional Park (see also: Afan Forest Park [254], Dare Valley Park [268] and Parc Slip [277]). From the visitor centre (Access point) built on the former colliery site (parking charge), a range of routes can be followed, from easy valley-bottom trails to more challenging paths among the wooded hills. At the south-east of the forest, on a prominent hill, are the remains of Twmbarlwm motte and bailey castle, built during the medieval period, most likely on top of an earlier Iron Age hill fort.

An alternative access option, and an unusual one in Britain, is a 11km restricted-access road (charge) through the forest known as Cwmcarn Forest Drive. Its barriers limit cars to a maximum of 120 at any one time. Seven car parks along the drive allow visitors to stop off and explore different parts of the forest, its various facilities (e.g. picnic sites and play areas), and wonderful panoramic views. For instance, car park P7 (ST2378192898 | ///implore.banks.positives) provides an ideal starting point to explore Twmbarlwm hill fort. All the sites can also be reached on foot from the visitor centre.

DAN-Y-GRAIG

RISCA, CAERPHILLY MAP REF 321

Ownership: Private: Gwent Wildlife Trust and Others **Designations:** **Area:** 5ha	**Forest type:** broadleaved **Forest location:** ST233905 **Explorer Map:** 152 **Ease of access:** Moderate	**Access point:** ST2351590523 ///soaks.caramel.outbound NP11 6ER

Growing on the lower eastern slopes of Mynydd Machen (362m), this small ash woodland regenerated naturally when the old Danygraig copper works closed. Active during the nineteenth century, the works were built on top of lead mines dating back to the Roman period. A maintained path runs through the wood, 1ha of which is managed as a nature reserve, including a section of boardwalk next to a small pond (a relic of the copper works). Look for slow worms in sunny glades, and watch dragonflies and damselflies hunting over the pond. The show of woodland flowers is beautiful in spring. Enjoy views across the valley to the peak of Twmbarlwm (419m) and Cwmcarn Forest [320]. The entrance to the reserve (Access point) is sandwiched between two bungalows. A car can be parked nearby along Danygraig Road.

COED WEN

PENHOW, NEWPORT MAP REF 322

Ownership: Public: Natural Resources Wales **Designations:** NNR, SSSI **Area:** 16ha	**Forest type:** broadleaved **Forest location:** ST417900 **Explorer Map:** 152 **Ease of access:** Moderate	**Access point:** ST4158389663 ///repeating.tricky.procures NP26 3AA

Coed Wen forms part of the Penhow Woods NNR and SSSI, which also includes a group of woods on the hill to the east, but none of which are accessible to the public. Its canopy of ash, small-leaved lime, wild cherry and wych elm allows ample sunlight to reach the lime-rich woodland soil, which harbours hosts of wildflowers. Coed Wen is a good site to see the rare native wild daffodil (*cennin pedr* in Cymraeg), which is more diminutive than the common garden cultivar, with paler petals and deep yellow trumpets. Also in spring, look for bluebell, green hellebore, lesser celandine and primrose, while the rare bird's-nest orchid can be seen in June. Among the many woodland birds to spot, bullfinch, redstart, tree pipit and pied flycatcher are particularly special. A relatively level path skirts the lower fringes of the wood from a lay-by (Access point), but some sections are steep, particularly if a circular route is followed around the circumference of this pretty wood.

A nuthatch calling sounds a little like a car alarm!

THICKET AND SLADE WOODS

LLANFIHANGEL NEAR ROGIET, MONMOUTHSHIRE MAP REF 323

Ownership: Public: Natural Resources Wales
Designations:
Area: 107ha

Forest type: mixed
Forest location: ST448890
Explorer Map: 154
Ease of access: Easy/Moderate

Access point: ST4483589416
///nobody.interlude.goad
NP26 3UF

On a small hill overlooking the confluence of the M4 and M48 (junction 23), these two adjoining woods provide plenty of informal access, offering good views across the Severn estuary and The Prince of Wales Bridge. Paths can be muddy, but that doesn't put off mountain bikers. Bluebells carpet the woodland in spring. Further down the hill from the Access point is Rogiet Poorland [325], and nearby the wildflower-rich meadow of Lower Minnetts Field nature reserve.

ST JULIAN'S WOOD

CHRISTCHURCH, NEWPORT MAP REF 324

Ownership: Public: Newport County Borough Council
Designations: LNR, SM
Area: 39ha

Forest type: broadleaved
Forest location: ST340893
Explorer Map: 152
Ease of access: Easy

Access point: ST3452989201
///festivity.scarf.zealous
NP18 1JJ

Overlooking the Usk Valley to the north, with the M4 motorway lying in the other direction, this large green space on the periphery of Newport was a deer park during medieval times. Today, it is much loved by local people. The woodland dates from at least the early 1600s and comprises pedunculate and sessile oak, ash, silver birch and sweet chestnut, with hazel, rowan and other small trees. A surprising diversity of wildlife can be found throughout the year. Resident birds include all three species of woodpecker, plus nuthatch and treecreeper. Skylarks sing above open grassy areas, where grass snake and slow worm bask. Traces of an Iron Age defensive camp with banks and ditches can be explored near the cemetery (ST3401089104 | ///wiping.rates.glass).

ROGIET POORLAND

LLANFIHANGEL NEAR ROGIET, MONMOUTHSHIRE MAP REF 325

Ownership: Private: Gwent Wildlife Trust
Designations:
Area: 1ha

Forest type: broadleaved
Forest location: ST452885
Explorer Map: 154
Ease of access: Easy/Moderate

Access point: ST4526988357
///bottom.removers.visual
NP26 3UF

When the sprawling 16,000ha Tredegar Estate was enclosed in 1855, local labourers were provided with an area of land to help support their subsistence, hence this site's name. During the twentieth century the poorland became derelict and gradually trees colonised the site. Now, ash, aspen and alder grow across the lime-rich soils, while open glades are rich in wildflowers and butterflies, including brown argus. Quiet visitors may be lucky enough to spot an adder. A short path circuits the wood, including some sections with steps. Further up the hill from the Access point lie Thicket and Slade Woods [323].

APPENDICES

Above: A young Scots pine tree grows by the Precipice Walk near Nannau [64].

Site Designations

Designation	Details	Number of sites in this guide
Heritage Park and Gardens (HPG)	A formal register by the Welsh Government's historic environment service, Cadw, of parks and gardens of special historic importance.	18
Important Plant Area (IPA)	A formal designation by conservation charity Plantlife for areas particularly rich with wildflowers, lower plants and fungi.	50
Local Nature Reserve (LNR)	A formal designation by Natural Resources Wales for areas of natural heritage that are locally important.	13
National Landscapes (NL)	Formerly known as Areas of Outstanding Natural Beauty (AONBs), there are five National Landscapes in Wales: Ynys Môn (Anglesey) and Llŷn Peninsula in the North-West, Bryniau Clwyd a Dyffryn Dyfrdwy (Clwydian Range and Dee Valley) in the North-East, Gower in the South-West, and Wye Valley in the South-East, which also extends into England.	41
National Nature Reserve (NNR)	A formal designation by Natural Resources Wales for areas with exceptional value and national importance for wildlife.	27
National Park (NP)	Areas designated for cultural heritage, landscape and nature. There are currently three national parks in Wales, covering 20 per cent of its land area: Bannau Brycheiniog/Brecon Beacons, Eryri/Snowdonia, and Pembrokeshire National Park (Parc Cenedlaethol Arfordir Penfro in Cymraeg). A fourth national park is currently under consideration for North-East Wales.	74
Nature Reserve	An informal designation by a landowner for an area of land considered to be valuable for wildlife, but without any formal designations.	
Scheduled Monument (SM)	A formal designation by Welsh Government's historic environment service, CADW, for nationally important sites spanning 8,000 years of human settlement.	73
Site of Special Scientific Interest (SSSI)	A formal designation by Natural Resources Wales for areas of land and water considered to best represent natural heritage, including flora, fauna and geology.	130
Special Area of Conservation (SAC)	Special Areas of Conservation are part of the UK's European site network of protected areas, formally designated under the EU's Habitats Regulations and still in force in Wales.	65
Special Protection Area (SPA)	A formal designation originating from the European Union, aiming to protect one or more rare, threatened or vulnerable bird species.	11

Cymraeg Terms

Tree names

Cymraeg	English
bedwen gyffredin	downy birch
criafolen/cerddinen	rowan
castanwydden felys/castanwydden bêr	sweet chestnut
cerddinen wen	common whitebeam
collen ffrengig	common walnut
derwen ddail digoes	sessile oak
derwen goesynnog	pedunculate oak
onnen	common ash
pinwydden yr Alban	Scots pine
ywen	yew
sbriwsen sitca/pyrwydden sitca	Sitka spruce
bedwen arian	silver birch

Wyndcliff Wood [315].

Rivers

Almost 90 different watercourses are mentioned in this guide. Many of these share the same name in Cymraeg and English. Where these differ, in the main text both names are given (English in brackets, or ordered Cymraeg/English). Those with different names in the two languages are also listed below.

Cymraeg	English
Alun	Alyn
Caeriw	Carew
Dyfi	Dovey
Dyfrdwy	Dee
Efyrnwy	Vyrnwy
Ewenni	Ewenny
Gwy	Wye
Hafren	Severn
Leithon	Ithon

Cymraeg	English
Llwchwr	Loughor
Mynwy	Monnow
Nedd Fechan	Neath
Ogwr	Ogmore
Rhaeadr Fawr	Aber Falls
Taf	Taff
Troddi	Trothy
Tywi	Towy
Wysg	Usk

A beech tree arches over the River Wye.

Geographic terms

Term	Description
aber	confluence of rivers
afon/afonydd	river/s
allt	wooded hillside, hillside, cliff, wood
boncyff	tree stem
bont	bridge
bryn	hill
cangen	tree branch
castell	castle
cefn	ridge
clogwyn	cliff
coed/coedwigoedd	wood/s
coeden	tree
cors	bog
craig	rock
cwm	valley
deilen/dail	leaf/leaves
dyffryn	valley
ffordd	road, way
ffridd	upland wood pasture
ffynnon	spring, well
garth	promontory, hill, enclosure
glan	riverbank, hillock
glyn	deep valley
gwaun	moorland
gwraidd/gwreiddiau	root/s

Term	Description
hafn	ravine, gorge
hafod	summer farmstead
hendre	winter farmstead
llannerch	clearing
llyn	lake
mawr	big
morfa/morfa heli	sea-marsh/salt marsh
mynydd	mountain, moorland
nant	brook, small valley
pant	hollow, valley
parc	park
penrhyn, pentir	promontory
pistyll	waterfall, spout
planhigfa	plantation of trees
plas	hall, mansion
pont	bridge
porth	gate/gateway, harbour
pren	tree
pwll	pool, pit
rhaeadr	waterfall
rhiw	hill, slope
rhos	moor, promontory
rhyd	ford
sir	county, shire
traeth	beach
ynys	island, river meadow
ystrad	valley, holm, river meadow

Glossary

ancient woodland	Forest area that has had continuous forest cover since 1600 in Wales and England (1750 in Scotland). *See also* semi-natural ancient woodland.
angiosperm	Flowering plants distinguished by having their reproductive organs carried in flowers, typically surrounded by petals and sepals. Following pollination, the closed ovary develops into a fruit. Includes all broadleaved trees. *See also* gymnosperm.
broadleaved	Trees belonging to the angiosperms that are usually deciduous (with a few exceptions in Britain such as holly).
bryophyte	A group of small non-vascular plants, including hornworts, liverworts and mosses. To the naked eye they can look plain and similar to one another, but under a hand lens or microscope their beauty and diversity can be appreciated.
champion trees	Individual tree specimens categorised as exceptional for their size (height or girth), great age, rarity or historical significance, and maintained in The Tree Register, a database of notable trees throughout Britain and Ireland collated by a registered charity.
climax (forest or ecosystem)	A plant community dominated by trees, representing the final stage of natural succession.
conifer	Trees belonging to the gymnosperms that are often cone-bearing and needle-leaved. With a few exceptions in Britain (e.g. larch) they are usually evergreen.
coppice/coppicing	To cut the stem of a tree to stimulate the production of multiple new stems to create a coppice stool.
deciduous	Tree species whose leaves fall at a certain time of the year (autumn in Britain). Usually broadleaved tree species, but also some conifer species. *See also* evergreen.

Natural regeneration of young conifer trees.

ecosystem	A community of organisms that interact with each other and the physical environment in which they exist.
epiphyte	A plant that grows upon another, such as on a tree branch, but which takes nothing from its host.
evergreen	Plants that do not shed all their leaves at once, therefore appearing green all year round. *See also* deciduous.
exclosure	An area from which undesirable mammals (especially herbivores such as deer) are excluded to help promote tree growth and natural regeneration.
forester	Professional embracing all aspects of the art, science and practice of creating, managing and conserving natural and humanmade forests.
ffridd	Term in Cymraeg for upland wood pasture, found on the boundary between intensively managed land at lower altitudes, and unenclosed hill and moorland.
gymnosperm	Division of flowering plants defined by their seeds being not enclosed in an ovary (cf. angiosperm). Includes all conifer species.
habit	The characteristic appearance, shape and form of growth of a plant species. Every individual plant's habit is a result of its genes in combination with the effects of the environment in which it grows.
krummholz	From the German 'knee timber', referring to stunted trees growing near the treeline at high altitudes, often found growing at the upper limit of montane forests.
lichen	A composite organism arising from a symbiotic partnership between algae or cyanobacteria, and fungi.
native	Species, subspecies or hybrids that established without the hand of humans more than 8,200 years ago (in Britain).
natural regeneration	The process by which ground is naturally replenished with trees that grow from seeds or suckers. Can apply to open ground or to clearings in existing forests.
rewilding	The restoration, usually at landscape scale, of ecosystems to allow nature to recover and then take care of itself through processes such as natural regeneration.
semi-natural ancient woodland	Forest area that retains native tree cover that has not been planted, although it may have been managed.
silviculture	Management of a forest, including its establishment, growth, composition, health and quality. Practised by a silviculturist (a discipline of forestry).
stand	Contiguous group of trees defined by the silviculture practised and/or by tree age, species composition or structure.
wood chain	The many stages or links between different elements of wood production and use, from the forest and the felling of its trees, and via road (or sometimes boat or rail) haulage, to timber, pulp or paper mills. End users in the wood chain include furniture makers, newspaper printers, firewood merchants, house builders, and many more.

Useful Information

Cadw	Welsh Government's historic environment service. cadw.gov.wales
Coetir Anian/ Cambrian Wildwood	A project centred in Mid Wales, working to restore habitats and species, connecting people with wildlife and wild places. cambrianwildwood.org
Coed Cadw/ The Woodland Trust	A membership charity planting trees and campaigning for protection of woodlands and trees across Wales. woodlandtrust.org.uk
Cyfoeth Naturiol Cymru/ National Resources Wales (NRW)	A Welsh Government-sponsored body with responsibility for caring for the natural resources of the country, including the public forest estate. naturalresources.wales
Llais y Goedwig	A company providing a coordinating role and membership for community woodland organisations in Wales. llaisygoedwig.org.uk
Lyme Disease Action	A charity dealing with Lyme disease throughout the UK, providing accredited information for the public, patients and health professionals. lymediseaseaction.org.uk
Mountain Rescue England and Wales	A charitable incorporated organisation that operates 13 volunteer groups across Wales offering help to lost and injured people in the outdoors. In an emergency, dial 999 and ask for Mountain Rescue. mountain.rescue.org.uk
National Trust Cymru	A membership charity protecting heritage in Wales. nationaltrust.org.uk
Phototrails	Online database of trails, particularly useful for those with a disability. phototrails.org
Plantlife	A charity campaigning to save wildflowers, plants and fungi, and responsible for the Important Plant Area (IPA) designation. plantlife.org.uk
RSPB Cymru	A charity caring and campaigning for wildlife, especially birds, across Wales. rspb.org.uk
Tree Register	A charity managing a database of more than 200,000 notable and champion trees in Britain. treeregister.org
Ymddiriedolaethau Natur Cymru/ Wildlife Trusts Wales	There are five Wildlife Trusts operating as independent charities in Wales: Gwent, Montgomeryshire, North Wales, Radnorshire and South and West Wales. wtwales.org

Further Reading

Atherton, I. et al. (2010). *Mosses and Liverworts of Britain and Ireland: A Field Guide*. British Bryological Society (BBS), London.

Barnard, P. C. (2011). *Royal Entomological Society Book of British Insects*. Wiley-Blackwell, Oxford.

Condry, W. (2009). *The Natural History of Wales: Collins New Naturalist Library, Book 66*. Collins, London.

Couzens, D., and Swash, A. (2017). *Britain's Mammals: A Field Guide to the Mammals of Britain and Ireland*. Princeton Wild Guides.

Hemery, G. (2023). *The Forest Guide Scotland: Copses, Woods and Forests*. Bloomsbury Wildlife, London.

Hemery, G., and Simblet, S. (2014). *The New Sylva: A Discourse of Forest and Orchard Trees for the Twenty-First Century*. Bloomsbury Publishing, London.

Holden, P., and Gregory, R. (2021). *RSPB Handbook of British Birds: Fifth Edition*. Bloomsbury Wildlife, London.

Johnson, O. (2015). *Collins British Tree Guide: Pocket Version*. Collins, London.

Langmuir, E. (2013). *Mountain Craft and Leadership: Fourth Edition*. Mountain Training Boards of England and Scotland. Cordee, UK.

Poland, J. (2020). *The Field Key to Winter Twigs: A Guide to Native and Planted Deciduous Trees, Shrubs and Woody Climbers (Xylophytes) Found in the British Isles*. Botanical Society of Britain and Ireland, St Albans.

Shields, C., and Ovenden, D. (2013). *Collins Fungi Guide*. Collins, London.

Sterry, P. (2008). *British Trees: A Photographic Guide to Every Common Species (Collins Complete Guide)*. Collins, London.

Streeter, D. et al. (2016). *Collins Wild Flower Guide*. Collins, London.

Below: Sessile oak trees overhang the Afon Tywi/River Towy at Gwenffrwd-Dinas [201].

REGIONAL MAPS

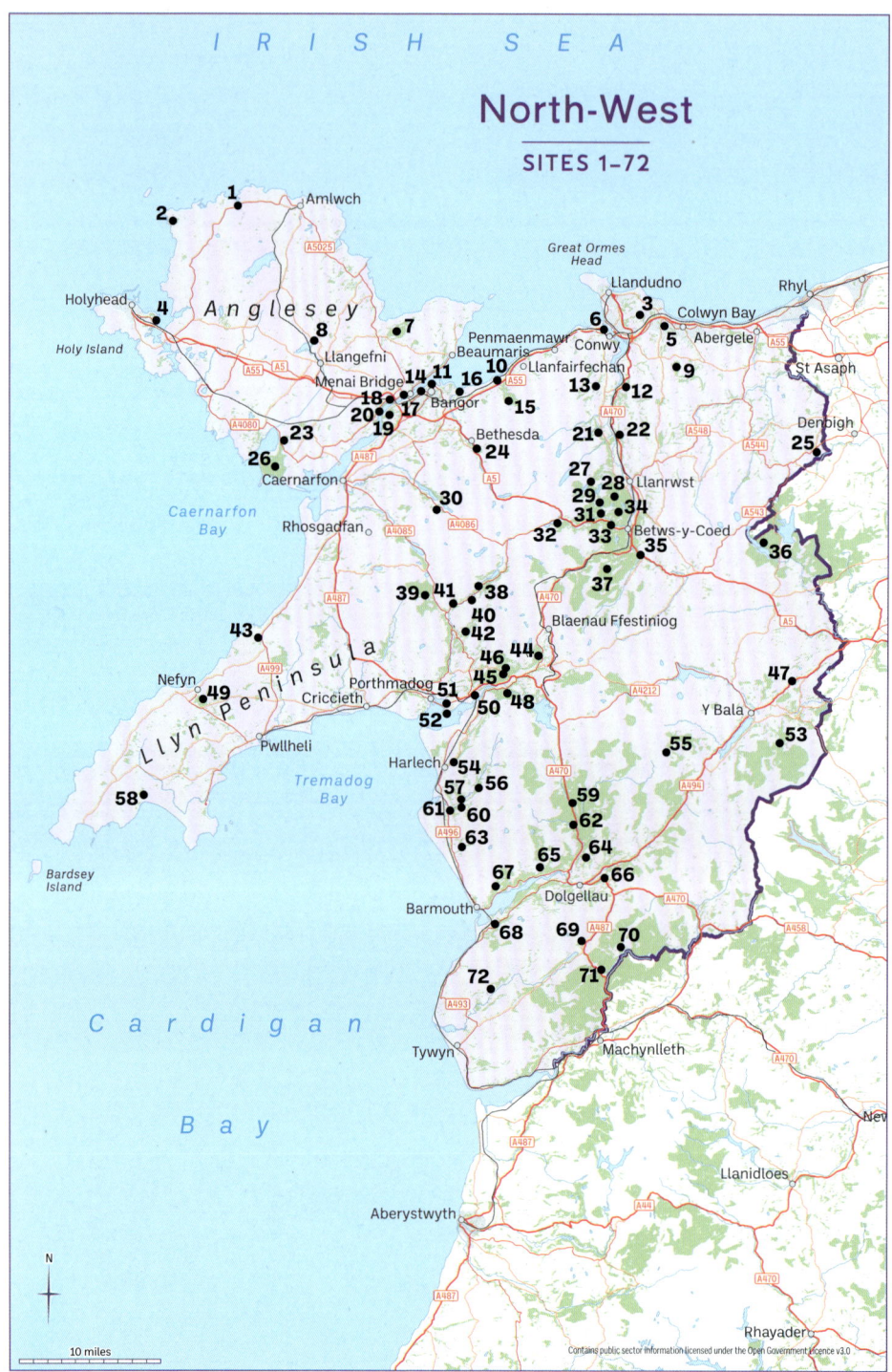

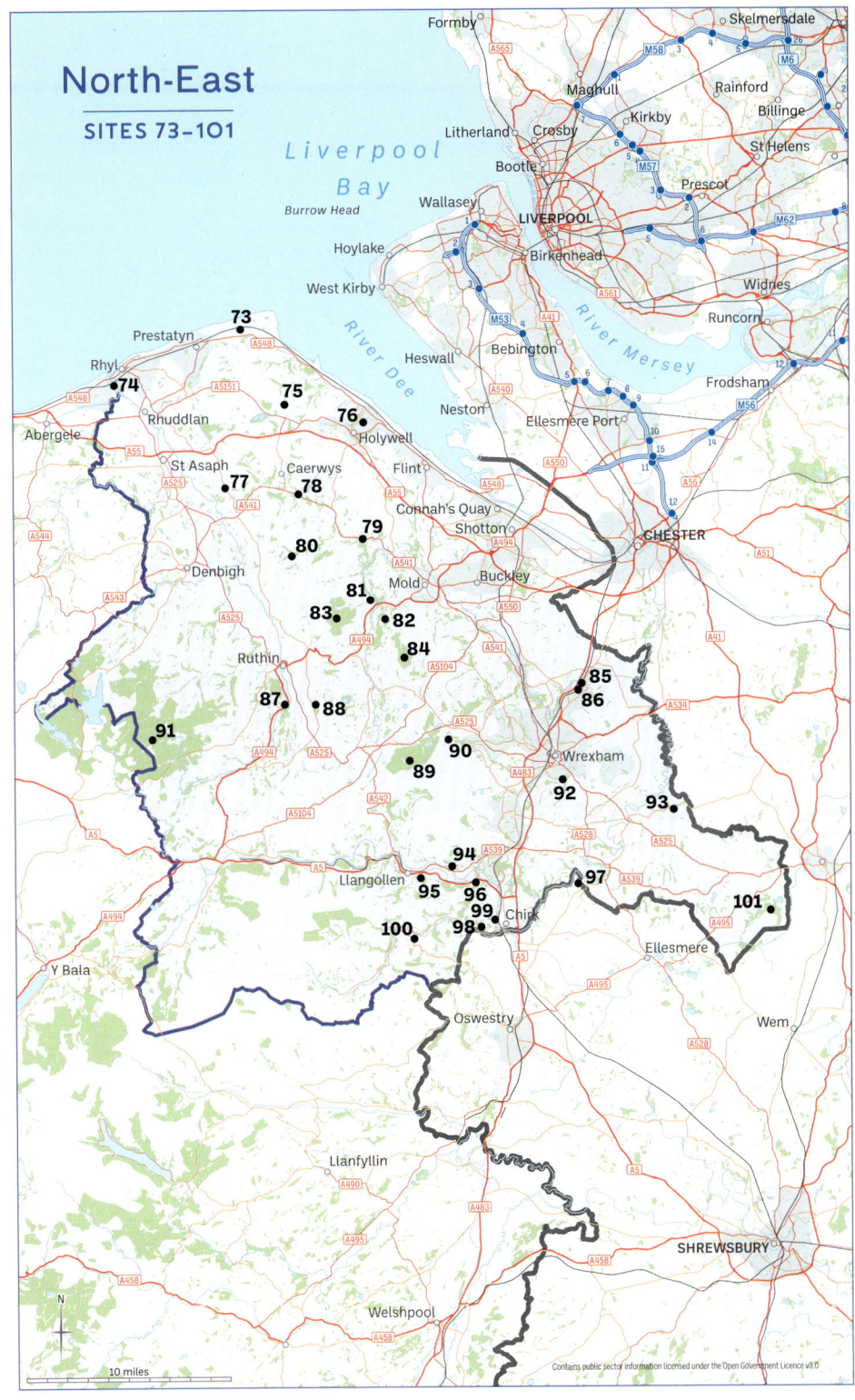

Central-East
SITES 136–200

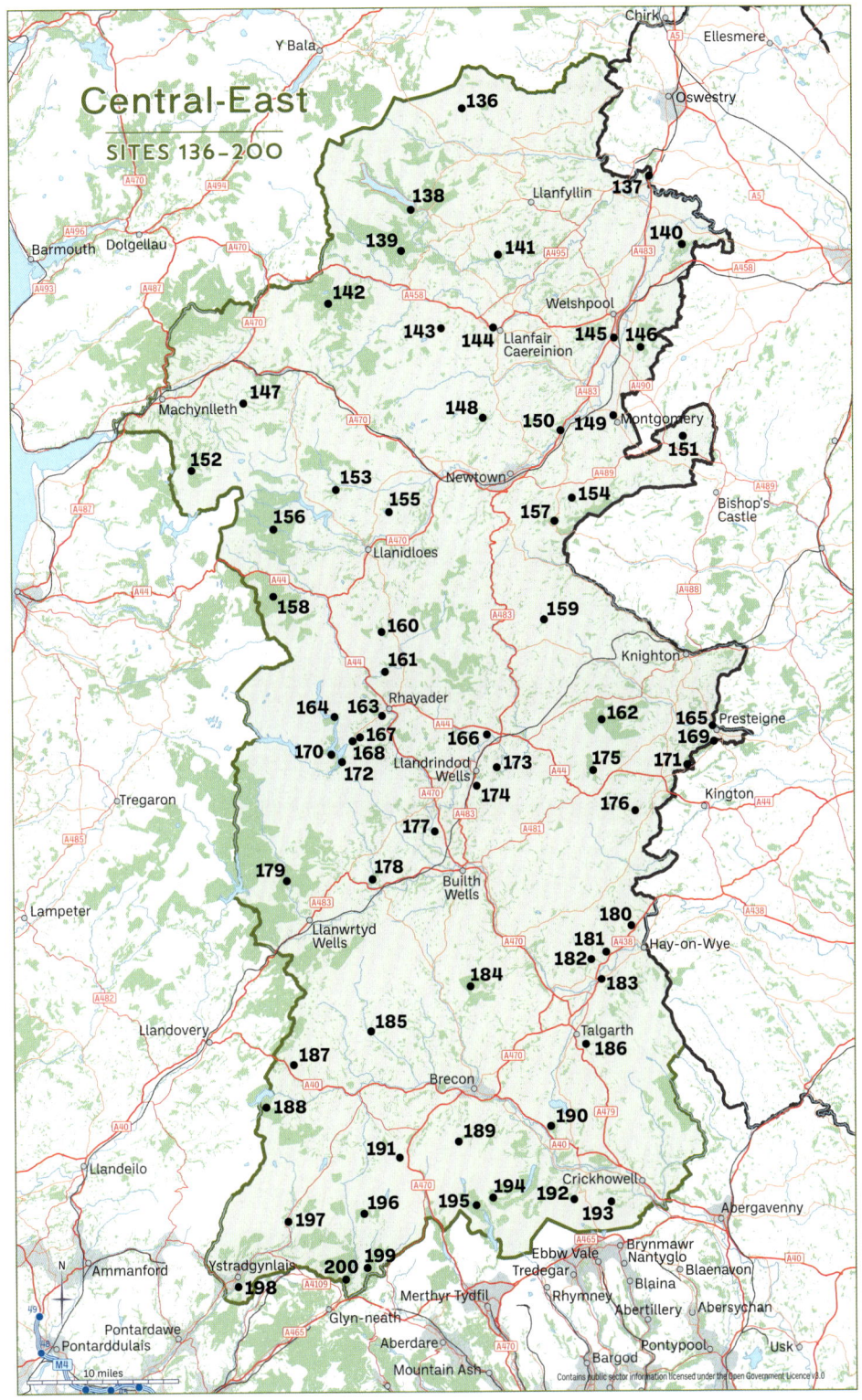

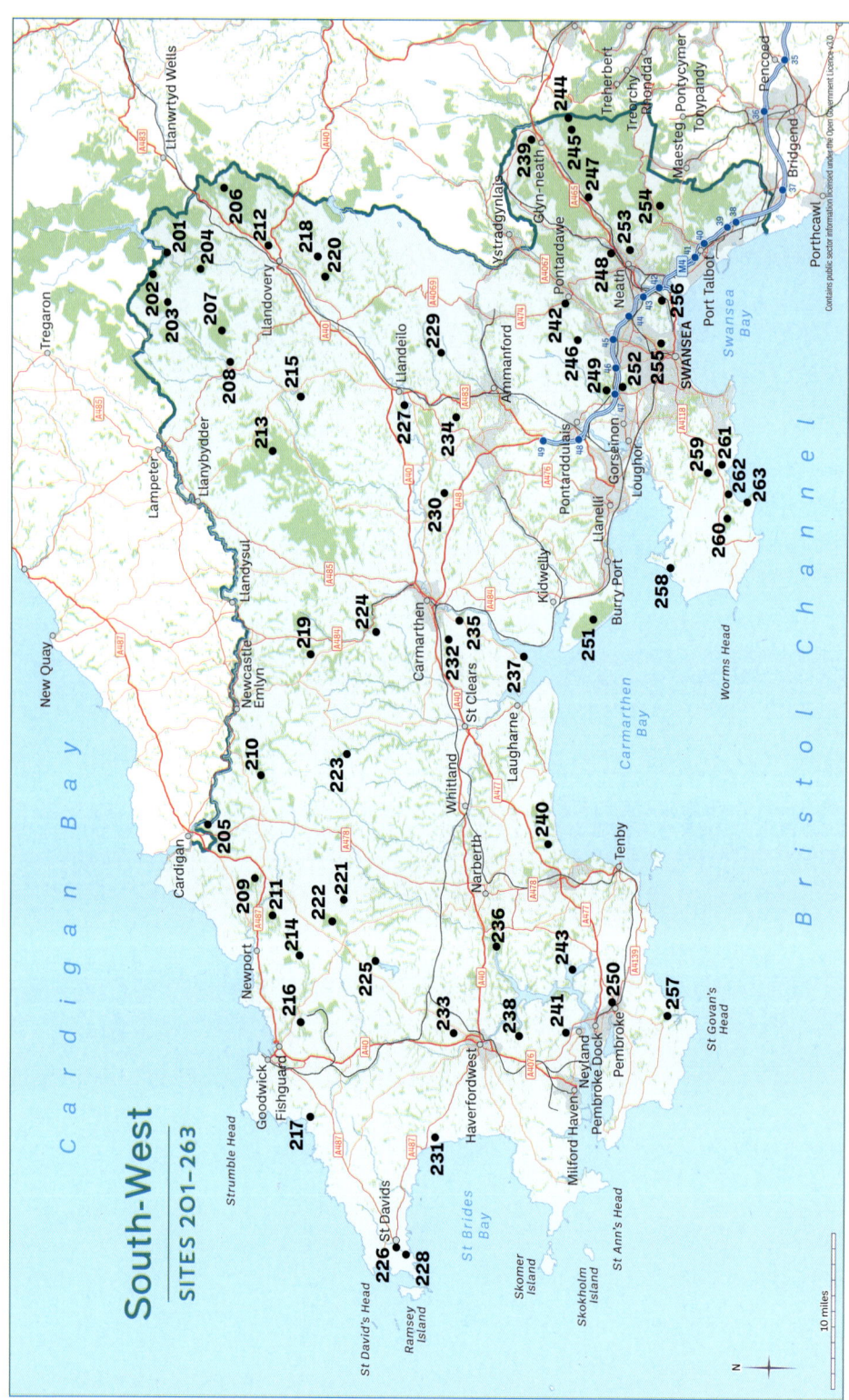

South-West
SITES 201–263

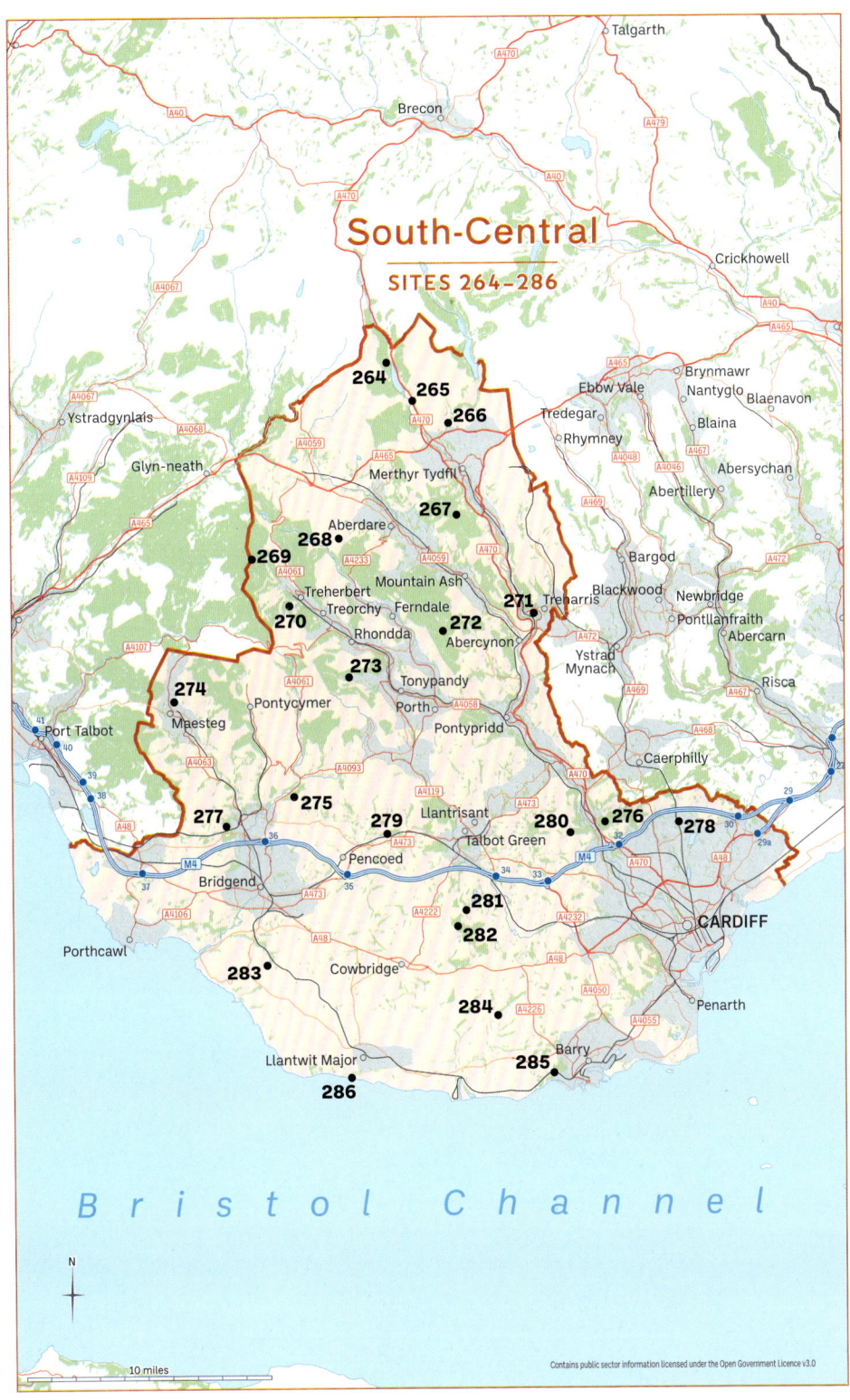

INDEX

Aber Artro Hall, Pentre Gwynnfrynn 70
Aber Mawr, Mathri/Mathry 183
Abercorris, Corris 77
Aberduna, Maeshafn 90
Aberhirnant, Bala 64
Abersychan Limestone Railway 256
Afan Forest Park, Pontrhydyfen 208
Afon Arto Valley 59, 66
Afon Cadair Falls, Minffordd 79
Afon Dyfrdwy 64, 84, 99, 102
Afon Gwy 9, 146, 255, 259
Afon Hafren 9, 136, 142, 259
Afon Mawddach 9, 74, 76
Afon Menai 37, 38, 41, 42, 43
Afon Taf 9, 220
Afon Teifi 9, 172
Afon Tywi 9, 126, 170, 190, 195
Afon Ystwyth 9, 17, 121, 122
Allt Boeth Pontarfynach/Devil's Bridge 119
Allt Crug Garn, Pennant 125
Allt Pencnwc, Ystrad Aeron 128
Allt Pontfaen – Coed Gelli-fawr, Pontfaen 179
Allt Rhongyr, Glyntawe 164
Allt Rhyd y Groes, Rhandirmwyn 170
Allt y Rhiw, Blackmill 223
Allt yr Esgair Woods, Llansantffraed 160
Alwen Forest, Cerrigydrudion 53
Anglesey, Isle of see Ynys Môn
Annell Aqueduct, Pumsaint 175
arboreta 35, 41, 70, 137, 139, 182, 192, 200–1
Arthog Bog, Fairbourne 78
Artist's Wood, Betws-y-Coed 51

Bae Abermaw/Barmouth Bay 62
Bae Abertawe/Swansea Bay 207, 208
Bae Ceredigion/Ceredigion Bay 9, 111, 114
Bae Colwyn/Colwyn Bay 35
Bae Tremadog/Tremadoc Bay 59, 65
Bailey Einon, Llandrindod 150
Banc-y-Castell, Goginan 12, 114
Bangor Forest Garden, Abergwyngregyn 36
Bangor Mountain 37
Bannau Brycheiniog 9, 10, 17, 130, 159, 160, 162, 164, 165, 166, 168, 235, 240–1, 247
Bargain Wood, Llandogo 256
Bat Sites SAC 54, 59, 65, 67
Bathesland Wood, Roch 193
bats 60, 63, 72, 75, 115, 139, 141, 145, 146, 150, 163, 172, 176, 177, 187, 198, 199, 206, 209, 211, 213, 214, 226, 246, 254, 262
Beacon Hill, Trellech 255
Beacons Way 161, 164, 165
Beaulieu Wood, Trefynwy/Monmouth 245
Beddgelert Forest, Beddgelerl 56
Berwyn Mountains 132
Big Pit National Coal Museum, Blaenavon 252
Big Pool Wood, Gronant 84

Black Covert, Llanilar 120
black grouse 53, 68, 94, 97, 137
Blackmill Woodlands, Blackmill 223
Black Mountains see Y Mynyddoedd Duon
Blaen y Glyn Falls, Pontsticill 165
Blaenant y Gwyddyl, Glyn-nedd/Glynneath 198
Blaenau Gwent 18, 234, 249, 255
Blaenavon Community Wood, Blaenavon 252
Blaenavon Industrial Landscape World Heritage Site 243, 252
Bodffordd, Bodffordd 35
Bodlondeb Woods, Conwy 32
Bodnant Garden, Llansanffraid Glan Conwy 35
Borth Forest, Borth 111
bothies, Grwyne Fawr, Partrishow 236
Moel Prysgau, Pontrhydfendigaid 127
Nant Rhys, Llangurig 144
Nant Syddion, Pontarfynach/Devil's Bridge 8, 113
Braichmelyn, Bethesda 45
Branches Fork Meadows, Pontnewynydd 260
Brechfa Forest Garden, Abergorlech 180
Brecon Becons see Bannau Brycheiniog
Brecon Mountain Railway 165
Breidden Forest, Crewgreen 136
Breidden Hills 131, 136
Bridgend 18, 216, 223
Broad Haven, Bosherston 209
Bronze Age sites 12, 72, 92, 143, 184, 210
Broom Wood, Mathri 183
Brycheiniog Forest, Llaneglwys 156
Bryn Arw, Bettws 239
Bryn Gwyn, Llanwddyn 134
Bryn Pydew, Bryn Pydew 32
Bryn y Fan 143
Brynau Wood, Castell-nedd/Neath 206
Bryniau Clwyd a Dyffryn Dyfrdwy 10 see also Clwydian Range; Dee Valley National Landscape
Brynna Woods, Brynna 226
bryophytes 10, 49, 50, 51, 53, 54, 56, 57, 58, 61, 65, 72, 73, 75, 79, 106, 108, 110, 115, 116–17, 118, 131, 150, 151, 171, 181, 190, 193, 219
Buckholt Wood, Trefynwy/Monmouth 238
Burfa Bank, Llanandras/Presteigne 149
butterflies and moths 32, 89, 90, 93, 94, 95, 133, 134, 140, 146, 177, 190, 199, 202, 204, 208, 219, 228, 229, 237, 242, 248, 255, 266
buzzard 56, 95, 119, 122, 135, 136, 137, 140, 141, 156, 174, 177, 179, 181, 186, 219, 236, 262

Bwlch Corog, Glaspwll 108
Bwlch Nant yr Arian Forest, Goginan 114
Bwllfa Colliery, Cwmdare 219

Cadair Idris 62, 73, 74, 76, 79, 81, 120
Cadair Ifan Goch 43
Cae Gaer fort, Esgair Ychion 12, 144
Caer Oleu Camp hill fort, Maenan 43
Caerphilly 234, 264–5
Caio Forest, Pumsaint 175
Cambrian Colliery, Tonypandy 222
Cambrian Mountains see Mynyddoedd Cambrian
Cambrian Way 56, 60, 113, 224
Cambrian Wildwood see Coetir Anian
camping 26
Canal Wood, Y Waun/Chirk 105
Cardiff 216, 224, 226
Carmarthenshire 168, 169, 170–1, 173–5, 181–2, 184, 186, 190, 191–2, 193, 194–5, 196, 204
Carmel, Pentre-Gwenlais 194
Carneddau Mountains 38, 39
Carngafallt, Rhaeadr Gwy/Rhayader 148
Castell Carn Dochan, Dolhendre 65
Castell Coch, Cardiff 224
Castell Cyfarthfa, Merthyr Tudful/Merthyr Tydfil 219
Castell Dinas Brân hill fort, Trevor 99, 101
Castell Dinefwr, Llandeilo 190
Castell Goetre hill fort, Llanbedr Pont Steffan 129
Castell Grogwynion, Pont-rhyd-y-groes 124
Castell Llansteffan, Llansteffan 196
Castell y Waun (Chirk Castle), Y Waun/Chirk 102
Castle Kinsey, Cleirwy 154
Castle Woods, Llandeilo 190
castles 13–14, 37, 65, 83, 102, 114, 143, 150, 154, 190, 195, 196, 211, 214, 219, 238, 264
Cefn Banog Ancient Village, Clocaenog 97
Cefn Cenarth, Pant-y-dwr 144–5
Cefnllys Castle, Llandrindod 150
Ceredigion 106–29
Ceredigion Coast Path 110, 123
Ceri Forest, Ceri/Kerry 143
Cerrig-y-Cledd standing stone, Abermaw 76
Charles Ackers Redwood Grove and the Naylor Pinetum, Tre'r llai/Leighton 138
chough 81, 123, 209
Cilcenni Dingle, Llowes 155
Cilgerran Castle, Cilgerran 14, 172
Cilgwyn Wood, Llangadog 184
Cleddau Estuary 197
Cleddon Falls, Llandogo 256
Cliff Wood, Y Barri/Barry 232
Clocaenog Forest, Clocaenog 97
Clwyd Forest, Llanferres 91

Index 283

Clwydian Range 10, 82, 83, 88, 90, 91, 92, 99, 101
Clydach Reservoir, Tylorstown 222
Clydach Vale, Tonypandy 222
Clywedog Gorge, Dolgellau 75
Clywedog Trail 98
Cnwch Wood, Rhaeadr Gwy/Rhayader 148
coal mining 14–15, 166, 169, 197, 198, 205, 208, 219–20, 222, 224, 225, 226, 230–1, 252
Coed Aberartro, Pentre Gwynfryn 70
Coed Allt Cefn Maesllan, Llanarth 128
Coed Allt-Fedw, Llanilar 122
Coed Allt Lan-las, Aberaeron 125
Coed Bryn Bras, Gwynedd 65
Coed Bryn-engan, Capel Curig 51
Coed Bryndansi, Dolwen 35
Coed Bwlch-derw, Beddgelert 56
Coed Cae Huddygl, Betws-y-Coed 50
Coed Cilgelynnen, Llanychaer 183
Coed Cilygroeslwyd, Pwll-glas 93
Coed Cnwch, Dolhendre 122
Coed Collfryn, Llwynmawr 103
Coed Cors-y-gedol, Tal-y-bont 72
Coed Craflwyn and Dinas Emrys, Beddgelert 57
Coed Craflwyn, Rowen 38
Coed Crafnant, Pentre Gwynfryn 66
Coed Cwm Clettwr, Tre'r-ddôl 110
Coed Cwm Einion, Furnace 108
Coed Cwm Tawel, Cynwyl Elfed 186
Coed-Cwnwr, Brynbuga/Usk 260
Coed Cymerau-isaf, Rhyd-y-Sarn 58
Coed Cyrnol, Porthaethwy/Menai Bridge 42
Coed Deri, Llanfair Caereinion 137
Coed Dolbebin, Pentre Gwynfryn 66
Coed Dolfriog, Nantmor 57
Coed Dolfudr, Dolhendre 65
Coed Dolgarrog, Dolgarrog 40–1
Coed Dolgoed, Ceredigion 122
Coed Dolwreiddiog, Pentre Gwynfryn 66
Coed Dolyronnen, Abercegir 138
Coed Elernion, Trefor 58
Coed Felenrhyd and Llennyrch, Maentwrog 61
Coed Ganllwyd, Ganllwyd 72
Coed Garnllwyd, Llancarfan 228
Coed Garth Gell, Bontddu 74
Coed Glaslyn, Four Crosses 135
Coed Gwernafon, Llawr y Glyn 141
Coed Gwraig, Tal-y-Coed 238
Coed Hafod y Bryn, Llanbedr 71
Coed Hafod y Llyn, Tan-y-Bwlch 59
Coed Hen Doeth, Llanberis 49
Coed Letter, Myddfai 184
Coed Llandegla, Pen-y-stryt 94
Coed Llanfairpwll, Llanfair PG 42
Coed Llangwyfan, Llangwyfan 88
Coed Llechwedd, Harlech 65
Coed Lletywalter, Pentre Gwynfryn 67
Coed Llwyn, Llanddunwyd/Welsh St Donats 227
Coed Llyn Mair, Tan-Y-Bwlch 60
Coed Maenarthur, Pont-rhyd-y-groes 124
Coed Mellte, Pontneddfechan 166–7
Coed Meyric Moel, Cwmbrân/Cwmbran 264
Coed Nant Brân, Llanfihangel Nant Brân 157

Coed Nedd, Pontneddfechan 167
Coed Nercwys, Maeshafn 92
Coed Nicholaston, Penrice 212
Coed Pen-y-Lan, Coed-y-paen 263
Coed Pen-y-maes, Treffynnon/Holywell 84
Coed Pencastell, Blaenwaun 186
Coed Pendugwm, Pontrobert 132
Coed Penglais, Aberystwyth 114
Coed Penglanowen, Aberystwyth 118–19
Coed Pwll-y-blawd, Cadole 89
Coed Rheidol, Pontarfynach/Devil's Bridge 115
Coed Shed, Goes 45
Coed Simdde Lwyd, Pontarfynach/Devil's Bridge 118
Coed Sylfaen, Abermaw/Barmouth 76
Coed Taf Fawr, Merthyr Tudful/Merthyr Tydfil 218
Coed Tan-yr-allt, Maenan 43
Coed Tre-Gynon Iron Age enclosure, Pontfaen 179
Coed Wen, Cwmbrân/Cwmbran 265
Coed Wern Ddu, Llanllwch 193
Coed y Bedw, Pentyrch 226
Coed y Bont, Pontrhyfendigaid 122–3
Coed y Brenin Forest Park, Dolgellau 68–9
Coed y Bwl, Castle-upon-Alun 228
Coed y Cerrig, Stanton 237
Coed y Cwm, Aberystwyth 112
Coed y Dinas, Y Trallwng/Welshpool 137
Coed y Felin, Hendre 87
Coed y Felin, Llys-faen/Lisvane 226
Coed y Foel, Llandysul 129
Coed y Garreg, Chwitffordd/Whitford 85
Coed y Parc, Park Mill 12, 210
Coed-yr-allt, Gwynedd 56
Coed yr Arch, Pontarfynach/Devil's Bridge 120
Coed Ysgubor-wen, Llanegryn 81
Coedlan Llanfairpwll, Llanfair PG 42
Coedmore, Cilgerran 172
Coedwigoedd Glaw Celtaidd Cymru 54
Coedydd Aber, Abergwyngregyn 39
Coedydd Dyffryn Ffestiniog, Tan-y-Bwlch 60
Coedydd Maentwrog, Tan-y- Bwlch 60
Coedydd Nedd a Mellte, Pontneddfechan 166
Coetir Anian 107, 108
Colby Woodland Garden, Amroth 198
conservation/restoration projects 11, 81, 107, 108, 119, 126–7, 142, 157, 171, 175, 177, 180, 189, 192, 209, 223, 230–1, 240–1
Conwy 16, 28, 32, 34, 35, 38, 40, 43, 45, 47–9, 50–3
Conwy Valley 43
Coppicewood College, Felindre Farchog 177
Corndon Hill 140
Cors Abercamlo, Y Groes/Crossgates 147
Cors Bodgynydd, Betws-y-Coed 49
Cors y Llyn, Newbridge-on-Wye 153
Cors-y-Sarnau, Sarnau 61

Craig Cerrig Gleisiad, Aberhonddu/Brecon 160–1
Craig Gwladus, Aberdulais 205
Craig Pysgotwr 171
Craig y Cilau, Llangatwg/Llangattock 162–163
Craig y Rhiwarth, Glyntawe 164
Croes Robert Wood, Trellech 253
Crychan Forest, Llanwrtyd Wells 174–5
Crymlyn Bog and Pant y Sais, Port Tennant 208
Cuckoo Wood, Trellech 256
Curley Oak, Wentwood Forest, Newport 262
Cwm Berwyn Forest, Tregaron 125
Cwm Byddog, Cleirwy/Clyro 154
Cwm Claisfer, Llangynidr 162
Cwm Clydach, Clydach 203
Cwm Coel, Rhaeadr Gwy/Rhayader 149
Cwm Colhuw, Llanilltud Fawr/Llantwit Major 229
Cwm Doethie – Mynydd Mallaen SAC 171
Cwm-Du Glen and Glanrhyd Plantation, Pontardawe 199
Cwm Luest, Blaenrhondda 221
Cwm Melin-cwrt, Melin-cwrt/Melincourt 204
Cwm Rhaeadr, Rhandirmwyn 173
Cwm Saerbren, Treherbert 220
Cwm Sere, Aberhonddu/Brecon 159
Cwm Taf Fechan, Merthyr Tudful/Merthyr Tydfil 219
Cwm yr Esgob 148
Cwmcarn Forest, Crosskeys 265
cycling/cycling routes 25, 33, 56, 64, 68, 97, 105, 114, 122, 127, 129, 141, 143, 144, 145, 149, 154, 165, 166, 173, 174, 186, 203, 208, 209, 220, 224, 249, 260, 264
Cymraeg/Welsh language 12, 16–17, 53

Dan-y-Graig, Risca 265
Dare Valley Park, Cwmdare 219
Dark Sky Discovery Sites 114, 120, 123, 125, 159
Darren Fach, Cefn-coed-y-cymmer 218
Darren Fawr, Merthyr Tudful/Merthyr Tydfil 218
Ddol Uchaf, Afon-wen/Afonwen 87
Dee Estuary 84, 85
Dee Valley National Landscape 10, 83, 99, 101
Denbighshire 16, 82, 84, 86, 88, 91, 93, 94, 97, 99
Deri-Fach, Y Fenni/Abergavenny 242
Devil's Gorge, Cilcain 89
Dinas Emrys, Beddgelert 13, 57
disabled access 26
diseases, tree 25, 35, 45, 165, 187, 194, 202, 203, 205, 206, 228, 252
dogs 25–6
Dolaucothi, Pumsaint 176
Dolforwyn Woods, Aber-miwl/Abermule 140
dormouse 75, 87, 110, 140, 154, 155, 156, 157, 177, 179, 226, 237, 248, 253, 258, 262

Douglas fir 8, 15, 44, 51, 52, 80, 122, 134, 147, 148, 153, 155, 173, 178, 183, 184, 187, 236, 256
Dyffryn Gardens, Vale of Glamorgan 200
Dyfi Estuary 109, 110
Dyfi Forest, Aberllefenni 80
Dyfnant Forest, Pont Llogel 135

Eagles Nest, Wyndcliff Woods, Monmouthshire 259, 261
Ebbw Vale Central Valley, Glynebwy/ Ebbw Vale 249
Elan Valley 145, 147, 148, 151
Elidir Trail 167
Epynt Way 159, 174
Erddig Country Park, Wrecsam/ Wrexham 98
Eryri National Park 9, 29, 37, 41, 42, 47, 50, 52, 56, 65, 76, 80, 120, 201
Esgair Ychion, Llangurig 144
Esgyrn Bottom, Llanychaer 183

Fan y Big 165
Fenn's Wood, Bettisfield 103
ferns see mosses
Ffestiniog Railway 60, 71
Fforest Fawr, Tongwynlais 224
Fforest Fawr UNESCO Global Geopark 218
Ffos Anoddun/Fairy Glen, Betws-y-Coed 53
Ffridd Wood, Trefaldwyn/Montgomery 140
Ffynnon Saint, Rhiw 70
Ffynone and Cilgwyn Woods, Capel Newydd/Newchapel 178
First World War 15, 97, 129, 199, 204
fish 131, 137, 144, 147, 172, 178
Flintshire 82, 84–5, 87, 89, 92
flycatcher, pied 39, 49, 54, 58, 63, 66, 67, 72, 74, 87, 89, 93, 107, 109, 110, 118, 119, 132, 134, 139, 140, 141, 144, 148, 150, 153, 154, 157, 170, 171, 172, 176, 179, 181, 185, 190, 197, 198, 204, 236, 237, 249, 265
spotted 42, 84, 95, 109, 147, 162, 172, 185, 249
forest management 15–16, 36, 51, 119, 185
Forest Stewardship Council (FSC) 15
Forest Wood, Cilgerran 172
Forestry Commission 15, 17, 29, 45, 97, 126, 156, 211, 227, 230–1
Fron Wood, Llowes 155
Fronderw Wood, Ceri/Kerry 142
fungi 33, 39, 40, 49, 102, 118–19, 135, 140, 146, 154, 203, 238, 243, 253, 255, 262

Gallt y Tlodion, Llanymddyfri/ Llandovery 181
Garn Boduan, Nefyn 62
Garth Bank, Garth 153
Garth Dingle, Llowes 155
geography of Wales 8–10
Gethin Woodland Park, Abercanaid 220
Gilfach Nature Reserve, Rhaeadr Gwy/ Rhayader 146

Gilfach y Dwn Fawr hill fort, Pontrhydfendigaid 123
Glamorgan see Merthyr Tydfil; Rhondda Cynon Taf
Glan Faenol, Bangor 43
Glanmor Fach, Aberystwyth 110
Glanrhyd House, Neath Port Talbot 199
Glasbury Cutting, Y Clas-ar-Wy/ Glasbury 156
Glasfynydd Forest, Trecastell/ Trecastle 159
Glen Morfa, Rhyl 84
Glynaeron Forest, Rosebush 185
Glyndŵr, Owain 73
Gnoll Estate Country Park, Castell-nedd/Neath 206
Goat Field Arboretum, Powys 137, 200
Gogerddan Wood, Penrhyn-coch 112
Goitre Coed Fach, Quaker's Yard 220
gold mining 13, 169, 175, 176
Gorlech Valley 180
goshawk 69, 97, 134, 135, 136, 137, 174, 185, 208, 253
Gower Ash Woods SAC 210, 212, 215
Gower Peninsula 10, 168, 212, 214
Graig Wyllt, Graig-fechan 94
Green Castle Woods, Llangain 195
Gregynog, Tregynon 139
grey wagtail 102, 110, 121, 132, 146, 155, 159, 165, 167, 198, 219, 236
Gro Wood, Rhaeadr Gwy/Rhayader 151
Grogwynion, Tynygraig 122
Grywne Fawr 236
Gwaith Powdwr, Penrhyndeudraeth 63
Gwenffrwd-Dinas, Rhandirmwyn 170
Gwili Heritage Railway 186
Gwydir Forest, Betws-y-Coed 52
the Gwyllt, Portmeirion 64
Gwynedd 16, 18, 28, 36–7, 38–9, 40, 43, 45, 49, 54, 56–81, 90

Hafna, Llanrwst 48
Hafod, Cwmystwyth 121
Hafod Elwy Moor, Cerrigydrudion 53
Hafod-y-Llan, Bethania 56
Hafren Forest, Llanidloes 142
Halfway Forest, Pontsenni/ Sennybridge 159
Harlech Castle, Harlech 65
Hendai medieval farmstead, Niwbwrch/Newborough 45
Hengwm Forest, Forge 141
Hensol Forest, Llanddunwyd/Welsh St Donats 227
hides, bird/wildlife 40, 44, 63, 84, 102, 109, 114, 134, 137, 181, 225
Highmeadow Woods, Trefynwy/ Monmouth 246–7
hill forts 12, 34, 37, 39, 43, 57, 62, 83, 88, 99, 114, 119, 122, 123, 124, 129, 130, 140, 149, 186, 238, 264
Holyland Wood, Penfro/Pembroke 205
Hook Wood, Maddox Moor 197
horse riding 25, 114, 129, 135, 143, 174, 186
Horseshoe Trail 16

Industrial Revolution 14–15, 224, 235, 252
Irfon Forest, Llanwrtyd 154
Iron Age sites 37, 39, 43, 119, 122, 123, 124, 129, 133, 140, 149, 179, 186, 229, 266
ironworks 14–15, 162, 166, 204, 220, 224, 252

Kerry Ridgeway Path 143
kestrel 136, 249
Kilvey Hill, Abertawe 207
kingfisher 40, 89, 102, 109, 137, 155, 157, 164, 165, 167, 199, 219, 236, 236
Knockmandown Wood, Y Barri/Barry 232
Knolton Wood, Owrtyn/Overton 102
the Kymin, Trefynwy/Monmouth 245, 247

Lady Park Wood, Trefynwy/Monmouth 246, 247
lakes, Ceredigion 127
 Conwy, 47, 49, 52, 53, 73
 Flintshire, 92
 Gwynedd, 8, 18, 56, 64
 Neath Port Talbot, 202, 203
 Powys, 134, 137, 143
 Ynys Môns/Anglesey, 29, 31, 35, 44
larch 15, 35, 43, 62, 118, 185, 187, 202, 203
Lasgarn Wood, Abersychan 258
Lawr yr Afon, Cemaes 30
Leat, Pumsaint 175
Lewis Wood, Wrecsam/Wrexham 98
lichens see bryophytes; mosses
limestone quarrying 83, 87, 95, 162, 166, 194, 197, 210, 224, 256, 258
Little Milford Wood, Maddox Moor 197
liverworts see bryophytes; mosses
lizards and newts 87, 204, 248, 266
Llais y Goedwig 18
Llanbrynmair Forest, Llangadfan 137
Llandyfeisant Church, Llandeilo 190
Llanerchaeron, Aberaeron 125
Llanerchi Wood, Rhaeadr Gwy/ Rhayader 149
Llanfihangel Cefnllys Church, Llandrindod 150
Llanfoist Wood, Llan-ffwyst/Llanfoist 243
Llangollen Canal 99, 103
Llanhaylow Wood, Gladestry 153
Llantwit Beach, Vale of Glamorgan 229
Llanymynech Rocks, Llanymynech 133
Llennyrch, Maentwrog 18, 28, 54, 58, 61
Llyn Crafnant, Trefriw 47
Llyn Fach, Hirwaun 202
Llyn Efyrnwy/Lake Vyrnwy 134
Llyn Parc Mawr, Niwbwrch/ Newborough 44
Llŷn Peninsula 10, 58, 62
Llyn Ty'n y Mynydd 49
Llyn y Fydlyn 31
Llynfi Valley 223
Llys y frân Reservoir, Llys-y-frân 187
Loggerheads Country Park, Codole 89
Long Wood, Llanbedr Pont Steffan/ Lampeter 129
Loughor, Llandyfan 191

Index **285**

Maen Madoc menhir, Ystradfellte 165
Maes y Gaer Camp hill fort, Abergwyngregyn 39
Maes y Pant, Gresford 93
Manor Wood, The Narth 253
Marford Quarry, Marford 92
Margaret's Wood, Whitebrook 252
Marteg Valley 146
Mawddach Valley 74
Meirionnydd Oakwoods IPA, Gwynedd 54, 58, 59, 61, 63, 65, 67, 72
Merthyr Tydfil 216, 218–19, 220
Migneint-Arenig-Dduallt, Dolhendre 65
military activity 26, 151, 174, 186, 199, 204, 232
Mill Wood, Penrice 211
Mill Wood, Y Barri/Barry 232
Minera Quarry, Minera 95
mining 14, 48, 50, 52, 80, 87, 89, 92, 95, 112, 113, 124, 141, 166, 167, 169, 172, 175, 176, 197, 198, 205, 208, 219–20, 222, 224, 225, 226, 230-1, 252, 256, 265
Ministry of Defence 17, 204
Minwear Forest, Arberth/Narberth 196
Moel Arthur hill fort, Llangwyfan 88
Monmouthshire 15, 234, 236–49, 250, 252–4, 255, 256, 258–9, 260–1, 263, 266
mosses, ferns, lichens and liverworts 28, 39, 43, 48, 50, 51, 52, 56, 57, 60, 61, 65, 66, 67, 74, 76, 77, 79, 102, 103, 115, 116–17, 144–5, 146, 147, 154, 161, 162, 166, 167, 170, 171, 172, 176, 179, 181, 183, 193, 204, 206, 242, 248, 249, 260
see also bryophytes
Myherin Forest, Pontarfynach/Devil's Bridge 112–13
Mynydd Bwllfa, Tonypandy 222
Mynydd Du, Partrishow 236
Mynydd Llwydiarth 34
Mynydd Machen 265
Mynydd Mallaen 171
Mynydd Preseli 178, 186, 187
Mynydd Ty-isaf 220
Mynyddoedd Cambrian 9, 108, 110, 112, 114, 120, 126, 131, 141, 145, 171, 178

Nannau, Llanfachreth 73
Nant Camllwydrew 226
Nant Gwynant 57
Nant Llygad Llwchwr 191
Nant Melin, Pumsaint 171
Nant-y Bedd forest garden, Partrishow 236
Nantgwynant, Bethania 56
Nantporth, Bangor 38
Nantygronw, Cwmduad 184
Nash Wood, Llanandras/Presteigne 148
National Botanic Garden of Wales, Porthyrhyd 192
National Landscapes (NLs) 9–10
National Nature Reserves (NNRs) 12
National Trust (Cymru) 17

Natural Resources Wales 7, 15, 17, 26, 231
Neath Port Talbot 168, 198, 199, 202–4, 205, 206, 208
Neolithic sites 179, 210
Newborough Forest, Niwbwrch/ Newborough 45
Newport 234, 262, 265, 266
nightjar 49, 63, 74, 103, 109, 136, 137, 159, 202, 207, 255
North Wales Path 32

Oak Wood Walk, Gilfach NR 146
Offa's Dyke 13, 82–3, 88, 91, 94, 99, 102, 138, 245
Old Warren Hill, Aberystwyth 119
orchids 32, 92, 93, 94, 112, 133, 193, 204, 210, 222, 228, 237, 248, 265
osprey 109, 142, 199
otter 75, 109, 115, 121, 131, 134, 144, 150, 155, 157, 165, 167, 172, 198, 206, 209
owl, barn 249
  long-eared 103, 137
  tawny 32, 66, 93, 95, 119, 123, 177, 190, 219
ownership and custodians 7, 17–18
Oxwich Castle, Swansea 215
Oxwich National Nature Reserve 212
Oxwich Wood, Oxwich 215

Pant Da, Capel Bangor 118
Pant-teg Wood, Felindre Farchog 177
Pantmaenog Forest, Rosebush 186
Parc Cenedlaethol Arfordir Penfro 168
Parc Coedwigaeth Pen-pych, Blaenrhondda 221
Parc le Breos Cwm and Coed y Parc, Parkmill 210–11
Parc Slip, Tondu 225
Parc Teifi Forest Garden, Aberteifi/ Cardigan 129
Pembrey Forest, Pen-bre/Pembrey 204
Pembrokeshire 18, 168, 169, 172, 177–9, 183, 185–6, 187, 191, 193, 196, 197, 198–9, 202, 205, 209
Pen Coed-Foel hill fort, Llandysul 129
Pen-y-castell Wood, Trefeglwys 143
Pen y Coed, Llangollen 99
Pen y Cymoedd, Hirwaun 203
Pen y Gaer hill fort, Trevor 99
Pen-yr-allt Wood, Marthi183
Penbont Woods, Rhaeadr Gwy/ Rhayader 147
Penderi Cliffs, Llanrhystud 123
Pendinas Reservoir, Wrecsam/ Wrexham 94
Pengelli Forest, Felindre Farchog 177
Penhow Woods, Penhow 265
Penllergare Forest, Penllergare 205
Penllergare Valley Woods, Penllergare 206
Pennant Melangell Church, Pont Llogel 135
Pennard Castle, Swansea 214
Pennard Cliff and Northill Woods, Parkmill 214

Penrhos Coastal Park, Caergybi/ Holyhead 33
Penrhyn Castle, Garth 37
Penrice Castle, Swansea 14, 211
Pentraeth Forest, Pentraeth 34
Pentre-Gwenlais turlough, Pentre-Gwenlais 194
Pentre Ifan dolmen, Felindre Farchog 12, 179
Pentrosfa Mire, Llandrindod 150
Pentwyn Farm, Pen-Twyn 248
Penycloddiau hill fort, Llangwyfan 83, 88
peregrine falcon 11, 41, 95, 123, 134, 137, 161, 163, 171, 185, 207, 220, 229
petrified forests 111, 183
Piercefield Woods, Cas-gwent/ Chepstow 261
pine, lodgepole 10, 15, 45, 62, 91, 125, 126–7, 250
pine marten 11, 115, 119, 121, 159
Pisgah Quarry, Froncysyllte 101
Pistill Cain waterfall, Dolgellau 68
Pistyll Rhaeadr, Llanrhaeadr-ym-Mochnant 132
plantations, impact of 20th century 15, 188–9
Plantations of Ancient Woodland Sites (PAWS) 54, 129, 184, 188–9
Plas Newydd, Llanfair PG 41
Plas-y-gors, Ystradfellte 165
Pontrhydfendigaid Trail 15
Porth Clais, Porthclais 191
Porthkerry Woods, Y Barri/Barry 232
Portmeirion village, Gwynedd 64
Powys 16, 80, 130–67
prehistoric sites 12, 37, 39, 41, 62, 72, 76, 85, 88, 92, 97, 111, 119, 122, 123, 124, 129, 133, 140, 141, 142, 143, 149, 153, 179, 186, 205, 208, 210, 220, 238, 264, 266
Preseli Hills see Mynydd Preseli
Priory Wood, Bettws Newydd 254
Prisk Wood, Pen-twyn 249
Programme for the Endorsement of Forest Certification (PEFC) 15
Pumlumon Fawr 141
Punchbowl, Llan-ffwyst/Llanfoist 250
Pwll-y-Wrach, Talgarth 157
Pwllycrochan Woods, Bae Colwyn/ Colwyn Bay 34
Pysgodlyn Mawr 227

Radnor Forest, Bleddfa 145
Radnor Valley 148
rainforest, temperate 7, 8, 10, 18, 28, 49–51, 53, 54, 56, 57, 58, 60, 61, 65, 66–7, 70–6, 79, 81, 84, 87, 106, 108, 110, 115, 118, 131, 141, 166–7, 170–1, 176–7, 179, 198, 204, 223
raised bogs 103, 183
rare insect, bird and animal species 28, 68, 95, 109, 119, 132, 134, 178, 191, 202, 213, 220, 227, 261
  see also dormouse; red squirrel

rare trees, plants and fungi 8, 11, 17, 32, 38, 39, 40, 48, 54, 61, 72, 87, 93, 108, 112, 115, 132, 139, 141, 151, 154, 155, 157, 161, 162, 171, 180, 193, 199, 202, 204, 210, 212-13, 218, 232, 237, 258, 265
see also orchids; rainforests
red kite 11, 81, 107, 114, 119, 122, 140, 148, 156, 159, 171, 186, 236
red squirrel 10, 11, 29, 30, 33, 34, 35, 41, 44, 45, 97, 125, 126-7, 154
redstart 39, 54, 58, 63, 66, 90, 95, 107, 109, 118, 132, 134, 139, 141, 144, 150, 153, 163, 170, 171, 176, 177, 190, 197, 198, 204, 236, 237
redwood, coast 138, 180, 199
giant 8, 35, 118, 134, 139, 180, 200
Rhaeadr Ddu, Ganllwyd 72
Rhaeadr Mawddach, Dolgellau 68
Rhaeadr Myherin, Pontarfynach/Devil's Bridge 112, 144
Rhaeadr Peiran, Cwmystwyth 121
Rhaeadr Y Graig Lwyd, Betws-y-Coed 53
Rhayader Tunnel and Embankment, Rhaeadr Gwy/Rhayader 145
Rheidol Valley, Pontarfynach/Devil's Bridge 115, 118, 119
the Rhinogs 66, 78
Rhondda Cynon Taf 216, 219, 220, 221-2, 226, 230
Rhondda Fach 222, 230
Rhondda Fawr 221, 230
Rogiet Poorland, Llanfihangel near Rogiet 266
Roman Camp, Garth 37
Romans/Roman sites 12-13, 14, 37, 57, 120, 135, 140, 144, 165, 175, 176, 191, 222, 265
Roundton Hill, Churchstoke 141
RSPB (Cymru) 18

safety 27
St Davids, Tyddewi/St Davids 187
St Gwynno Forest, Tylorstown 222
St Julian's Wood, Christchurch 266
Second World War 63, 67, 151, 186, 204
service tree species 199, 232
Sgwd Gwladus, Pontneddfechan 167
Sgwd yr Eira falls, Pontneddfechan 166
Silent Valley, Cwm 255
Sites of Special Scientific Interest (SSSI) 12, 42, 60, 81, 110, 125, 132, 135, 137, 153, 183, 187, 197, 208, 218, 219, 224, 248, 258, 265
snakes, adder 89, 199, 208, 248, 266
grass 87, 266
Snowdon 54, 56, 62, 73
Snowdonia National Park see Eryri National Park
sparrowhawks 177, 179, 253
Spinnies Aberogwen, Bangor 40
Spirit of Llynfi, Maesteg 223
Springfield Farm NR, Brynbuga/Usk 260
spruce, Sitka 10, 15, 33, 49, 53, 62, 64, 70, 91, 94, 125, 126, 137, 141, 147, 154, 159, 175, 185, 186, 187, 202, 203, 222

Norway 45, 119, 153, 154, 184, 211, 220, 256, 263
Stackpole Estate, Bosherston 209
stone monuments, prehistoric 76, 97, 208
Strata Florida Abbey, Pontrhydfendigaid 123, 127
Strawberry Cottage Wood, Stanton 236
Stump Up For Trees 12, 235, 239, 240-1
Sugar Loaf see Y Fâl
Swallow Falls, Betws-y-Coed 50
Swansea 168, 203, 205, 206, 207, 208, 210-15
Sychryd cascades, Pontneddfechan 166

Taf Fechan Forest, Pontsticill 165
Taff Trail 164, 218, 219, 220
Tal y Fan 38
Talley Abbey, Talyllychau 182
Talley Woods, Talyllychau/Talley 182
Talybont Forest, Talybont-on-Usk 164
Talyllychau Valley 182
Tan-y-pistyll, Llanrhaeadr-ym-Mochnant 132
tanning industry 14, 54
Tarren y Bwllfa 219
Thicket and Slade Woods, Llanfihangel near Rogiet 266
ticks and lyme disease 27
Torfaen 234, 252, 256, 260, 264
Traeth Glaslyn, Minffordd 63
Traeth Ynys Y Fydlyn, Llanfairynghornwy 31
Tranch Wood, Pontnewynydd 260
Trawsgoed Fort, Llanilar 12, 120
Trevor Hall Wood, Trevor 99
Twmbarlwm hill fort and castle, Caerphilly 264
Twyn y Bridallt camp, Tylorstown 222
Twyn-y-Gaer hill fort, Stanton 237
Ty Canol, Felindre Farchog 179
Tyddewi/St Davids Peninsula 191
Tylcau Hill, Moelfre City 144
Tŷ Mawr Wybrant, Penmachno 53
Ty'n y Bedw, Tynygraig 122
Tywi Forest, Pontrhydfendigaid 126-7
Tywi Valley 173

UNESCO Dyfi Biosphere Reserve 80, 108
upland hill farming 11-12, 240-1
Usk Valley 162, 242, 250, 260, 266

Vale of Clwyd 82, 86, 91
Vale of Ffestiniog 60, 61
Vale of Glamorgan 15, 16, 216, 227-9, 230
Vale of Llangollen 101
Vale of Trefaldwyn/Montgomery 140
Valeways Millennium Heritage Railway 229
Valle Crucis Abbey, Llangollen 99
Valleys Regional Park 208, 219, 225, 231, 264
Vaynol Park, Bangor 43
Velindre Wood, Llys-y-frân 187
Vincent Wildlife Trust 11

Wales Coast Path 196, 204, 210, 214
Wales National Forest 52
Warren Wood, Maesyfed/New Radnor 152
waterfalls 53, 68, 72, 102, 112, 115, 121, 132, 137, 141, 149, 152, 157, 159, 164, 165, 166-7, 173, 178, 198, 199, 204, 206, 221, 222, 256
Wat's Dyke Way 13, 83, 102
Waun Fach 236
Waun Las NNR, Porthyrhyd 192
Welsh Highland Railway 56
Welshpool and Llanfair Light Railway 137
Wentwood Forest, Llanvair Discoed 262-3
The Wern, Mitchel Troy 248
Wern Plemys, Ystradgynlais 166
West Williamston, West Williamston 202
Westfield Pill, Neyland 199
wetland birds 32, 38, 40, 42, 43, 63, 84, 109, 148, 183, 190, 196, 197, 199, 202, 210
whitebeam 161, 163, 164
English 8, 258, 261
Leys 8, 218
Menai 8, 38
Whiteford Burrows, Llanmadoc 210
Wildlife Trusts 18
Withy Beds, Llanandras/Presteigne 147
Withybush Woods, Hwlffordd/Haverfordwest 193
wood warbler 54, 60, 66, 67, 74, 107, 109, 110, 118, 132, 139, 140, 153, 162, 171, 177, 179, 181, 186, 198, 204, 237
Woodland Assurance Standard, UK 15
Woodland Trust (Coed Cadw) 18
woodpecker 53, 171, 181, 190
great spotted 74, 102, 128, 135, 219, 253, 255, 264, 266
green 92, 229, 262, 266
lesser spotted 54, 65, 106, 109, 148, 162-3, 228, 266
Wrexham 81, 92, 93, 94-5, 98, 101, 102-5
Wurthymp Wood, Worthenbury 98
Wye Valley 10, 234, 247, 248, 253, 256, 261
Wye Valley Walk 146, 247, 249, 255, 256
Wyndcliff Wood, Cas-gwent/Chepstow 258-9
Wysis Way 247

Y Fâl 235, 239, 242, 255
Y Graig, Tremeirchion 86
Y Mynyddoedd Duon 9, 131, 159, 160, 162, 236, 238, 239, 244, 247, 260
Ynys-hir, Eglwys-fach 109
Ynys Môn 8, 10, 16, 28, 29, 30-1, 33, 34-5, 37, 41-2, 44, 45
Ysgyryd Fawr, Y Fenni/Abergavenny 244
Ystrad Gwyn, Minffordd 79
Ystwyth Gorge, Pont-rhyd-y-groes 124

Index 287

ACKNOWLEDGEMENTS

Completing this guide took considerable time, many thousands of miles of travel and months of planning, even before writing could begin. A very special word of thanks to my Book Patrons (listed in the front) for their generous support, and to Book Sponsors: Carolyn Thorne, David Mitchell, Dušan Farrington, Jamie Balfour-Paul, James Collett, James Ogilvie, Jim Chiazzese and Sarah Watkinson. I am also indebted to the many members of the public who suggested sites for inclusion.

A carved Welsh dragon at Porthkerry Woods [285].

For help with fieldwork and research, I am grateful to Alison Sherriff, Andy Stott, Christopher Williams, Dagmar Vesely, Daniel Ackerley, Eliza Kaczynska-Nay, Huw Denman, James Collett, John Deakin, John Healey, Keith Powell, Lisa Handcock, Nick Thomas, Peter Jones, Robert Penn, Rod Waterfield, Simon Lloyd, Steve Gurney and Sally Malam, and Wendy Necar. Also to the following organisations: National Trust, Royal Forestry Society, Small Woods Association, and Stump Up For Trees.

To the wonderful team at Bloomsbury, especially Amy Hodkin, Katy Roper and Jim Martin, supported by Austin Taylor, Elizabeth Peters, Lucy Beevor, John Plumer, Kate Inskip and Delyth Davies.

Finally, to my wife Jane and family, for their unerring support and encouragement, and great company during fieldwork.

ABOUT THE AUTHOR

Gabriel Hemery is a writer, photographer and forest scientist. He co-founded and is chief executive of Sylva Foundation; a charity caring for forests across Britain.

This guidebook is Gabriel's second in a series featuring the copses, woods and forests of Britain published by Bloomsbury Wildlife. *The Forest Guide: Scotland* was published in 2023, and *The Forest Guide: England* is forthcoming. He is also the author of *The New Sylva*, which was published by Bloomsbury in 2014 (republished in 2021).

To discover more about the Forest Guide series, visit: gabrielhemery.com/forest-guide. Readers can enjoy free access to an interactive online map to all sites featured in this guide. Use the password *copsewoodforest* to gain full access.